T0135536

Extreme Nonlinear Optics with Spatially Controlled Light Fields

seit 1558

Dissertation

zur Erlangung des akademischen Grades

doctor rerum naturalium (Dr. rer. nat.)

vorgelegt dem Rat der Physikalisch-Astronomischen Fakultät

der Friedrich-Schiller-Universität Jena

von

Dipl.-Phys. CHRISTIAN KERN

geboren am 09.02.1982 in Oberwesel

Jena 2014

Bibliografische Information der Deutschen Nationalbibliothek

Die Deutsche Nationalbibliothek verzeichnet diese Publikation in der
Deutschen Nationalbibliografie; detaillierte bibliografische Daten sind
im Internet über http://dnb.d-nb.de abrufbar.

Zugl.: Jena, Univ., Diss., 2014

ISBN 978-3-8325-3817-0

Logos Verlag Berlin GmbH
Comeniushof, Gubener Str. 47,
10243 Berlin
Tel.: +49 (0)30 42 85 10 90
Fax: +49 (0)30 42 85 10 92
INTERNET: http://www.logos-verlag.de

List of Publications

Journal articles

- J. Lohbreier, S. Eyring, R. Spitzenpfeil, C. Kern, M. Weger, and C. Spielmann, "Maximizing the brilliance of high-order harmonics in a gas jet", New Journal of Physics **11**, 023916 (2009)

- R. Spitzenpfeil, S. Eyring, C. Kern, C. Ott, J. Lohbreier, J. Henneberger, N. Franke, S. Jung, D. Walter, M. Weger, C. Winterfeldt, T. Pfeifer, and C. Spielmann, "Enhancing the brilliance of high-harmonic generation", Applied Physics A **96**/1, 69–81 (2009)

- C. Kern, M. Zürch, J. Petschulat, T. Pertsch, B. Kley, T. Käsebier, U. Hübner, and C. Spielmann, "Comparison of femtosecond laser-induced damage on unstructured vs. nano-structured Au-targets", Applied Physics A **104**/1, 15–21 (2011)

- S. Eyring, C. Kern, M. Zürch, and C. Spielmann, "Improving high-order harmonic yield using wavefront-controlled ultrashort laser pulses", Optics Express **20**/5, 5601–5606 (2012)

- M. Zürch, C. Kern, P. Hansinger, A. Dreischuh, and C. Spielmann, "Strong-field physics with singular light beams", Nature Physics **8**/10 743–746 (2012)

- M. Zürch, C. Kern, and C. Spielmann, "XUV coherent diffraction imaging in reflection geometry with low numerical aperture", Optics Express **21**/18 21131–21147 (2013)

- C. Kern, M. Zürch, P. Hansinger, A. Dreischuh, and C. Spielmann "Extreme nonlinear optical processes with beams carrying orbital angular momentum", Proc. SPIE **8984**, Ultrafast Phenomena and Nanophotonics XVIII, 89841A (2014)

Conference contributions

- S. Eyring, C. Kern, N. Franke, S. Jung, C. Spielmann, R. Spitzenpfeil, C. Ott, J. Lohbreier, J. Henneberger, D. Walter, M. Weger, C. Winterfeldt, and T. Pfeifer, "Enhancing the brilliance of high-harmonic radiation", poster at 441. WE-Heraeus-Seminar on Ultrafast X-ray methods for studying transient electronic structure and nuclear dynamics, Bad Honnef/Germany (2009)

- C. Kern, M. Zürch, S. Eyring, and C. Spielmann, "Limitations of ultrafast nonlinear nano-optics", poster at International Summer School in Ultrafast Nonlinear Optics SUSSP 66, Edinburgh/Scotland (2010)

- C. Kern, M. Zürch, and C. Spielmann, "Limitations of extreme nonlinear ultrafast nano-photonics", poster at 4th European Conference on Applications of Femtosecond Lasers in Materials Science & Nano and Photonics, Mauterndorf/Austria (2011)

- C. Kern, M. Zürch, B. Kley, T. Pertsch, and C. Spielmann, "Considerations and Requirements of Metallic Nanostructures for Plasmon-Enhanced High-Harmonic Generation", poster at Ultrafast Optics UFOIX, Davos/Switzerland (2013)

- C. Kern, M. Zürch, P. Hansinger, A. Dreischuh, and C. Spielmann, "Extreme Nonlinear Optical Processes with Beams Carrying Orbital Angular Momentum", oral presentation at European Conference on Lasers and Electro-Optics and the International Quantum Electronics Conference (CLEO/Europe-IQEC) München/Germany (2013)

Acknowledgements

A piece of work such as this is never the sole accomplishment of the person whose name is sported on the front cover. I thankfully acknowledge the invaluable contributions, material and immaterial, of the following people and institutions.

- Christian Spielmann for the years of mentoring. He has sparked interest in the projects forming this work. His expertise and inquisitiveness has been a guide through all phases.

- The Carl-Zeiss-Stiftung for enough interest and confidence in my work to secure financial and personal support for the largest part of my time spent on this thesis.

- My colleague Michael Zürch for co-working on many of the presented experiments. His There-I-fixed-it-approach has enabled great parts of the realisation of this work.

- Alexander Dreischuh for igniting and resolving the project that became part II of this thesis, and his exhaustive knowledge on the topic. Peter Hansinger for discussions and interpretation of the unrefined results.

- Jörg Petschulat for simulations and design discussions on all things concerning nano-antennas in general, and bow-ties in particular. Ernst-Bernhard Kley and Michael Banasch for realisation of the bow-tie sample. Thomas Käsebier for production of the gold layers, Christian Helgert for providing the nano-sphere samples, Uwe Hübner for providing the samples of nano-rectangles. Stefan Fasold and Matthias Falkner for the resonance measurements on the bow-ties. Hans-Jürgen Hempel for the occasional SEM session.

- Michael Damm, Wolfgang Ziegler and Burgard Beleites for support on all technical things in the laboratory and workplace. All colleagues, former and present, of the Quantenelektronik group—Stefan Eyring, Sebastian Jung, Nico Franke, Robert Spitzenpfeil, Michael Schnell, Andreas Hoffmann, Björn Landgraf, Maximilian Gräfe, Daniil Kartashov—for discussions, help, suggestions, and for being fun to be around.

- All my friends, in Jena and elsewhere; everyone who has played music with me.

- My parents, always curious for every single step in the process of developing and writing this thesis, with more generous patience than sometimes even I could muster.

- Britta, my dearest, without you there just would not be a point.

»There is a theory which states that if ever anyone discovers exactly what the Universe is for and why it is here, it will instantly disappear and be replaced by something even more bizarre and inexplicable.

There is another theory which states that this has already happened.«

DOUGLAS ADAMS

Abstract

The Generation of High-Harmonics (HHG) has been established as an indispensable tool in optical spectroscopy. This effect arises for instance upon illumination of a noble gas with sub-picosecond laser pulses at focussed intensities significantly greater than $10^{12}\,\mathrm{W/cm^2}$. HHG provides a coherent light source in the extreme ultraviolet (XUV) spectral region, which is of importance in inner shell photo-ionisation of many atoms and molecules. Additionally, it intrinsically features light fields with unique temporal properties. Even in its simplest realisation, XUV bursts of sub-femtosecond pulse lengths are released. More sophisticated schemes open the path to attosecond physics by offering single pulses of less than 100 as duration. The resulting large values of peak brilliance have made these lab-sized setups a light source complementing high average brilliance synchrotron facilities for the soft x-ray spectral regime.

The spatial phase of the laser pulse used for HHG is decisive for the yield and quality of the generated XUV light. This is due to its influence on the effect at the level of a single atom on the one hand, and the necessity of additive signal generation in the laser's propagation direction (phase matching) on the other. Controlled shaping of the phase, or adaptive shape finding, has the potential to increase obtained harmonic signals by several orders of magnitude. However, due to the complexity of the process, the dependence of the phase of the output XUV light on spatially controlled laser light is non-trivial.

Field enhancement by plasmonic nano-antennas has been claimed to boost local laser field strengths, at the site of the interaction, from initial intensities too low for HHG to sufficient values. This was done by inserting arrays of bow-tie-shaped antennas of $\sim 100\,\mathrm{nm}$ in length into the interaction region, close to the gas. As the far field High-Harmonic signal would then consist only of the coherent superposition of single point emitters in the hot spots of the antennas, this can be considered a phase-shaping process. However, its feasibility depends on the vulnerability of these nano-antennas to the still intense driving laser light. We show, by looking at a set of exemplary metallic structures, that the threshold fluence F_{th} of Laser-induced Damage (LID) is a greatly limiting factor for the proposed and tested schemes along these lines.

We separate pure material damage from LID on nanoscopic metallic particles. The factors that determine LID in the transition from bulk-like to nanoscopic are examined by evaluating laser shot sites on thin gold films. Furthermore, we find that the quality of the plasmonic resonance of nano-antenna samples makes these structures more sensitive to damage. This is verified by looking at non-resonant metallic spheres as well as explicitly resonant metallic rectangles of nearly equal dimensions. Finally, we look at lithographically produced bow-tie antenna arrays as they would be used for an experiment as described. Variations in fabrication allow a quantitative study of F_{th} in correlation with the exact resonance. Again we find that a higher quality of the resonance makes the most useful antennas more susceptible to LID.

In another experiment using phase shaped pulses we examine the transfer of an Optical Vortex (OV) into the XUV by HHG as an extremely nonlinear effect. An OV is a phase feature that incorporates helical phase patterns around a central dark (singular) region of zero field. Under certain conditions it can be shown that an OV introduces quanta of Orbital Angular Momentum (OAM) into the light field, allowing a whole new angle at XUV spectroscopy. We show, to our knowledge, the first example of a Vortex in the XUV spectral region that has been produced by imprinting the phase feature onto the fundamental light, and that has survived the extreme nonlinearity of an HHG process. We detect the OV in the XUV by evaluating spatial as well as phase features. An intuitive expectation that the charge of the resulting Vortex—or number of quanta of OAM—is multiplied by the harmonic order in the process cannot be confirmed.

Zusammenfassung

Die Erzeugung Hoher Harmonischer (High-Harmonic Generation, HHG) als Werkzeug ist aus der optischen Spektroskopie nicht mehr wegzudenken. Dieser Effekt entsteht etwa bei der Bestrahlung von Edelgasen mit Laserimpulsen mit einer Dauer von weniger als einer Picosekunde unter Fokussierung auf Intensitäten, die 10^{12} W/cm^2 deutlich überschreiten. Die HHG ermöglicht die Erzeugung von kohärentem Licht im extrem ultravioletten (XUV) Spektralbereich, welcher bei der Photoionisation der inneren Schalen vieler Atome und Moleküle von Bedeutung ist. Außerdem macht sie Lichtfelder mit einzigartigen zeitlichen Eigenschaften verfügbar. Selbst mit den einfachsten Realisierungsmöglichkeiten werden Lichtblitze im XUV erzeugt, welche intrinsisch Pulsdauern im Sub-Femtosekunden-Bereich besitzen. Mittels ausgefeilterer Methoden erschließt sich das Feld der Attosekunden-Physik, wo einzelne Pulse von weniger als 100 as Dauer benötigt werden. Die daraus resultierende hohe Spitzenbrillanz hat diese experimentellen Aufbauten, die bequem in ein Labor passen, zu einer Lichtquelle werden lassen, welche die enormen Durchschnittsbrillanzen von Synchrotron-Quellen im weichen Röntgenbereich gewinnbringend ergänzt.

Die räumliche Phase eines Laserimpulses, der für ein HHG-Experiment verwendet wird, bestimmt die Ausbeute und Qualität des erzeugten XUV-Lichtes. Dies ist einerseits durch ihren Einfluss auf den Effekt am einzelnen Atom begründet, andererseits aber auch durch die notwendigerweise additive Erzeugung von Signallicht in Propagationsrichtung des Lasers (Phasenanpassung). Eine kontrollierte oder adaptiv angeleitete Phasenformung erlaubt es, das gewonnene Harmonischen-Signal um mehrere Größenordnungen zu erhöhen. Durch die Komplexität des Prozesses ist jedoch die Abhängigkeit der Phase des XUV-Outputs von der räumlichen Kontrolle des Laserlichtes nichttrivial.

Eine Feldüberhöhung durch plasmonische Nanoantennen soll es ermöglichen, das lokale Laserfeld in der Wechselwirkungsregion zu verstärken, so dass aus Intensitäten, die ursprünglich unterhalb der Schwelle für HHG liegen, ausreichend hohe Intensitäten erzeugt werden können. Verwirklicht hat man dies durch das Einbringen von Arrays aus bow-tie-förmigen, etwa 100 nm großen Antennen in die Nähe des wechselwirkenden Gases. Da dann das Hohe-Harmonischen-Signal im Fernfeld nur aus der kohärenten Überlagerung von einzelnen Punktemittern — den "Hot Spots" der Antennen — besteht, kann man dies als einen phasenformenden Prozess betrachten. Jedoch hängt dessen Erfolg von der Empfindlichkeit dieser Nanoantennen gegenüber dem immer noch intensiven Laserlicht ab. Anhand verschiedener exemplarischer metallischer Strukturen zeigen wir, dass die Schwellfluenz F_{th} für Laser-induzierte Beschädigung (Laser-induced Damage, LID) ein stark limitierender Faktor von vorgeschlagenen und erprobten experimentellen Realisierungen ist, die auf diesem Konzept basieren.

Wir trennen dabei die Zerstörung des bloßen Materials von der LID nanoskopischer metallischer Partikel. Dabei werden die Faktoren, welche LID im Übergangsbereich zwischen ausgedehntem Festkörper und nanoskopischem Objekt bestimmen, durch die Auswertung von Laser-Einschüssen in dünnen Goldfolien untersucht. Darüber hinaus stellen wir fest, dass die Güte der plasmonischen Resonanz dafür verantwortlich ist, wie empfindlich Nanoantennen auf Laserbeschuss reagieren. Wir bestätigen dies, indem wir zum einen nichtresonante metallische Kugeln, und zum anderen explizit resonante metallische Nanorechtecke von vergleichbarer Größe betrachten. Schließlich untersuchen wir lithographisch hergestellte Bow-tie-Antennenarrays, wie sie für das beschriebene Experiment verwendet würden. Durch Abweichungen in der Fertigung ist es uns möglich, eine quantitative Untersuchung darüber durchzuführen, wie F_{th} mit der exakten Resonanz korreliert. Auch hier zeigen wir, dass eine höhere Güte der Resonanz die am besten nutzbaren Antennen anfälliger für LID macht.

In einem weiteren Experiment unter Verwendung phasengeformter Pulse untersuchen wir die Übertragung eines Optischen Vortex (OV) ins XUV, mittels HHG als Beispiel eines extrem nichtlinearen Effektes. Ein OV ist eine Phasenkonfiguration, welche aus einer spiralförmigen Abhängigkeit der Phase rund um eine zentrale Nullstelle (Singularität) des Feldes besteht. Unter bestimmten Voraussetzungen kann man zeigen, dass ein OV dem Lichtfeld Quanten an Bahndrehimpuls hinzufügt, was dem Feld der XUV-Spektroskopie völlig neue Möglichkeiten eröffnet. Wir zeigen das unseres Wissens nach erste Beispiel eines Vortex im XUV-Spektralbereich, welcher erzeugt wurde, indem die Phasenform dem Fundamentalen aufgeprägt wird, und welcher die extreme Nichtlinearität des HHG-Prozesses überlebt. Der Nachweis des OV im XUV geschieht durch Auswertung von räumlichen sowie phasenbezogenen Merkmalen. Die intuitive Annahme, dass sich die Ladung des entstandenen Vortex — bzw. die Anzahl an Bahndrehimpulsquanten — mit der Harmonischenordnung multipliziert, kann nicht bestätigt werden.

Contents

Chapter 1

Introduction to the Subject Matter

The desire for higher spatial and temporal resolution for ever more accurate measurements of physical systems is a driving force behind the development of controlled light sources to shorter wavelengths, higher fluxes, shorter stroboscopic bursts of radiation. Neglecting the wavelength scales of x-rays and beyond—significant for strongest bound electron spectroscopy and structural evaluation of solids—we focus here on the range bridging the ultraviolet end of the visible spectrum and the soft x-ray regime, called the extreme ultraviolet (XUV). This is commonly regarded as the range from a few eV to 1 keV in photon energy, corresponding to a few hundreds of nanometres down to 1 nm in terms of wavelength. The energy of this type of radiation is of the order of inner shell electrons in many atoms, as well as intermediate binding energies in molecules, and is thus useful for probing these energetic states in various phases [1].

What is even more important is the intrinsic connection of an electromagnetic wave in the XUV, by the length of its optical cycle, to time scales in the attosecond regime ($1\,\mathrm{as} = 10^{-18}\,\mathrm{s}$). At these time scales, the fast electronic processes within matter become resolvable. So, if one is to achieve the production of light pulses in the XUV significantly shorter than a femtosecond ($1\,\mathrm{fs} = 1000\,\mathrm{as}$), one could take a 'snapshot' of the transient states of matter at different stages, making ultrafast processes directly visible [2,3].

One enormous breakthrough on the road to this goal came in the early 1990's, when Corkum and Kulander independently realised the significance of an effect that happened upon the ionisation of atoms in strong laser fields [4,5]. The availability of pulsed lasers—with a pulse duration of some tens of picoseconds and below—that offer very intense fields at their pulse peak had led to experiments to a degree of nonlinearity beyond the first few orders [6]. During the interaction of laser pulses of intensities greater than $10^{13}\,\mathrm{W/cm^2}$ with a (gaseous) nonlinear medium, the strength of the respective effect deviated from the nth-power scaling law inherent to $\chi^{(n)}$-type processes. The theoretical treatments mentioned above related the observed generation of 'high' harmonics to the classically described acceleration of an electron in the same strong electric field that was responsible for creating this free electron through ionisation from an atom. They furthermore succeeded in describing the characteristic quantities, such as the highest achievable 'cut-off' frequency, rather accurately even with this simple model. The subsequent research and application of this effect have become too large for even an inner-disciplinary complete overview, rapidly earning itself the capitalised version of its name, High-Harmonic Generation (HHG).

Already in its early stages, HHG enabled access to a coherent, laser-like source of XUV radiation [7]. Using more refined methods, mainly affecting the ionisation step, the second goal stated above was achieved, namely the temporal concentration of XUV light to the attosecond regime [8–10]. The physics at these time scales was established as the extreme end of 'ultrafast' physics, which however historically encompasses femtosecond phenomena that are even accessible by solid state lasers. Since 'ultrafast' is a superlative in itself, the term 'atto(second) science' is sometimes used when exploring real sub-femtosecond dynamics. In any case, what ensued was an abundance of experiments probing the dynamics of atoms and molecules in a whole new way [11,12], with sophisticated versions of the process itself leading to the insight into some of the most fundamental questions in physics [13].

This alone more than makes up for the rather low conversion efficiency of the HHG process [14,15], but still, for some experiments the brilliance or number of photons per time interval impinging on a sample is of relevance. Hence, the optimisation of HHG in terms of brilliance has been an ongoing topic in the community in general [16,17], and our group in particular [18–20]. The brilliance, or brightness, of a light source is the standard when comparing photon-producing elements, and is the benchmark especially when researching light-matter-interactions with low rates or cross-sections. HHG sources, as shall be seen, have the benefit of emitting radiation intrinsically in packets of sub-femtosecond duration, an advantage that is reflected by classifying sources in terms of *peak brilliance*, defined as

$$B_{\text{peak}} = \frac{N}{\Delta\tau \cdot \Delta\Omega \cdot A \cdot 0.1\% \, BW}. \tag{1.1}$$

Here, N is the number of photons, $\Delta\tau$ is the duration of the light pulse, $\Delta\Omega$ is a solid angle interval, and A is the size of the source area. $0.1\% \, BW$ denotes the percentage of the bandwidth $\Delta\nu/\nu$, in order to give a spectrally comparable number.

When temporal resolution is not of interest, and wherever the efficiency of detectors is a limiting factor, the sheer number of photons in a standardised time interval Δt, for a given geometry, is relevant. The *average brilliance*

$$B_{\text{av}} = B_{\text{peak}} \cdot \frac{\Delta\tau}{\Delta t} \tag{1.2}$$

can be used to quantify the average illumination of a sample during an experiment. Here, the extremely bright synchrotron sources are the go-to solution [21], whereas in peak brilliance experiments, HHG sources become competitive, considering their far less demanding setup [2,19]. Taking the properties of High-Harmonic XUV sources as a pulsed source as given, one could increase its flux or number of photons per second by increasing the repetition rate at which the HHG process takes place. Furthermore, efforts to produce High-Harmonics at higher repetition rates are driven by the desire to build frequency combs in the XUV [22], where the separation of spectral modes, or the finesse of the comb, is inversely proportional to the XUV pulse rate. This, however, places high requirements upon the driving light fields.

In the common schemes, lasers that are amplified to deliver the necessary intensities for the process are employed, and their shot repeatability is limited by the thermal load on the amplifying material to a few tens of kHz [23]. Newer generations of fibre lasers can crank the repetition rates up to some hundred kHz [24,25], while some unique, not run-of-the-mill laser systems have recently been shown to be able to push the conventional schemes up to 20 MHz [26]. All of these schemes employ a sequential approach to HHG, in that they *first* produce the intense light field, and then apply this intense field to the interaction region. Breaking up this sequential *modus operandi*, in some experiments the HHG interaction was relayed to an amplifying external build-up cavity [22], or was placed within the original oscillator cavity [27]. On the other hand, as it will later be described, the HHG process is very sensitive to the spatial phase of the fundamental field inside a small volume around the focus, where the interaction takes place. A lot of research in our group has been conducted to examine these subtle influences on phase matching, especially upon adaptive shaping of the spatial phase [18–20, 28]. The very aspect of adaptive shaping highlights the complexity of the involved parameters and their interplay. Even for the simple influential factors it is easier to run through the available parameters experimentally—for the phase shaping even automatised—and then to evaluate the effect through the observation of the resulting spectrum.

The focus of the present work is to explore High-Harmonic Generation as an extremely nonlinear process under conditions where the spatial phase is severely modified, either necessarily or intentionally. The distinctiveness of the two experimental projects that are documented here suggests that they should be arranged in two parts. In part I of this work, we shall examine the feasibility of a rather novel approach to boosting driving laser fields locally, at the site of a gas target for HHG [29]. Conceptually, this attempt makes use of the localisation of electromagnetic energy, through the use of plasmonic nano-antennas, to a spatial extent below the free-space diffraction limit. This concentration is limited to very small volumes, where the intensity of an impinging field can be enhanced to values sufficient for HHG. In contrast to conventional free focussing, High-Harmonic emission from gas in the vicinity of a large number of field-enhancing plasmonic devices is limited to certain nanoscopic regions *inside* the focus. This scheme can thus be considered to be a severe modification of the spatial shape of the light field driving HHG. Propagational phase matching, a backbone of efficient frequency conversion, is furthermore suppressed by the small axial extent of the enhanced field regions. Also, from a practical point of view, the intricate metallic structures are prone to Laser-induced Damage (LID) rendering them useless at external field intensities that enter the strong field regime. The result of a parametric study of LID of nano-antennas as they would be used for such an experiment will be presented.

Part II of this work deals with a more constructive, dedicated way of controlling the spatial phase of a light field that is used in an High-Harmonic experiment. A specific phase shape of peculiar properties is given by an Optical Vortex (OV). OVs are constructed from helical phase patterns that can be linked to light beams carrying Orbital Angular Momentum (OAM). This degree of freedom allows intriguing new possibilities in spectroscopy, as will be outlined in the theoretical section of part II. Presented here is experimental evidence that the phase pattern with all of its peculiarities can be transferred from the fundamental driving field over an High-Harmonic process. We thus provide a new technique for producing Optical Vortices in the XUV spectral region.

Both parts of this work begin with a chapter that introduces the fundamental ideas relevant to the respective topic. The second chapter of each part provides the necessary information on the experimental background, with some technical details outsourced to the appendices. Results from our performed experiments are then presented in a structured format. A short summary of our findings, as well as the context of these findings in respect to the research others have done on each topic, forms the concluding chapter of each of the two parts.

Part I

Limitations of Plasmon-Assisted High-Harmonic Generation

Chapter 2

Theory of Plasmon-Assisted High-Harmonic Generation

In 2008 an experiment was heralded that promised not only an easily aligned setup for a source of High-Harmonics at conventional oscillator rates of 75 MHz, but it also seemed to be a fundamental conjunction of two formerly separate fields in optical physics. Kim *et al.* [29] published High-Harmonic spectra that they had produced in the traditional strong field process by making use of field enhancing plasmonics in metallic nano-structures. While the near limitless spatial shaping of optical near fields in plasmonics [30] has led to a fantastic range of fundamental scientific as well as real world applications [31, 32], its combination with classical strong field effects is a very recent development [33, 34]. The details on the workings of this scheme shall be broken down in this chapter. A conceptual sketch of the experiment that led to the publication in [29] is shown in Fig. 2.1.

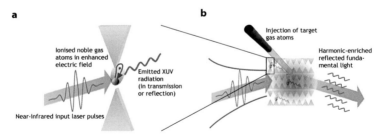

Figure 2.1: Scheme of High-Harmonic Generation aided by plasmonic field enhancement (after [29]). **a**, An incident, near-infrared ultrashort laser pulse excites a plasmon inside a bow-tie nano-antenna. Noble gas atoms, as commonly used in conventional HHG, experience the enhanced field between the tips of the triangles. **b**, When a large array of single bow-tie antennas is used, harmonics are generated at each 'hot spot', resulting in a directional beam in the far-field.

In the following years simulations that accounted for the conditions in that experiment elucidated some intriguing features of this concept. For example, an extension of the HHG cut-off energy over the free focus case, in the absence of a field shaping plasmonic structure, was theoretically obtained [35, 36], and a means of exploiting the spatially shaped fields for polarisation gating was proposed [37]. However, the experimental realisation of these conditions, including the present work, ran into trouble. While the spectral features seen in [29] were not perfectly reproduced in any work [38–40], some flatly contradicted Kim *et al.*'s findings, pin-pointing lines in spectra obtained in such an experiment to 4π incoherent line emission [40]. Meanwhile, Kim's group reported follow-up experiments that admittedly showed severe limitations to their previously published results [41]. Laser-induced Damage seemed to be an issue, as it was universally feared to be the case for this experiment walking a fine line between high and too-high fields.

Even before publication of these findings from different groups, this thesis aimed to explore the boundaries of the intriguing initial idea. The theoretical methods for this are founded in the Generation of High-Harmonics in noble gases, as well as the plasmonic response of metallic nano-structured objects to electromagnetic fields. From a practical point of view, the energy deposition of ultrashort laser pulses into metal targets of varying shape was to be explored, as well as the plasmonic characteristics of bow-tie antennas as they would be used for plasmon-enhanced HHG. A theoretical basis for all of the concepts that are important for part I of this thesis will be developed in this chapter.

2.1 'Conventional' High-Harmonic Generation

The 'simple man's' three-step model of High-Harmonic Generation (Fig. 2.2) has been explained in detail in many works, also by this group [1], and is, after two decades of existence, already considered 'textbook knowledge' within the community. It shall, as such, only be sketched briefly in this section, while giving a little more depth to the quantum-mechanical formulation of each step, as this is the basis for its extension to inhomogeneous fields, later in section 2.4.

A brief description of the three-step model

The first simple model that was to describe High-Harmonic Generation on a single atom was oriented on the classical experiment of focussing an intense laser pulse into a rare or noble gas [6]. It would have to account for the extremely high reachable photon energies that were produced, as well as the feature of equally intense harmonics over a large region called the 'plateau', and its high-energy cut-off (Fig. 2.3). The underlying premise of the model is that it is a 'semi-classical' description—in an adaptive manner, the fully

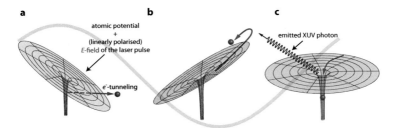

Figure 2.2: Three-step model of High-Harmonic Generation. **a**, Ionisation of an atom by a strong electromagnetic field, resulting in a free electron. **b**, Subsequent acceleration of the free electron in the field, reversal of direction after half an optical period. **c**, Return of the electron to the original ion, recombination under emission of a high energetic photon. All steps are described in more detail in the respective sections below.

quantum-mechanical process is treated at well-defined points, which mark the break down into separate steps, in the simplest, Newtonian and classical electromagnetic terms. The cornerstones, namely ionisation and recombination, however, rely on established quantum-mechanical models.

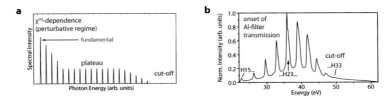

Figure 2.3: Schematic (**a**) and measurement (**b**) of a typical High-Harmonic spectrum, from argon.

In its original form, two approximations are made at different stages of the process:

- **Single active electron** (SAE) approximation: All correlation effects of many-electron systems are ignored, treating the atom as hydrogen-like. The implications of this only affect the ionisation and recombination steps, and are justified by using proper models for both.

- **Strong field approximation** (SFA): While the ionisation step that starts the whole process is sensitive to the temporal dependence of the electric field, as shall be described below, at the 'time of birth' t_0 when the electron enters the laser field, its state is completely reset, as if the external field had not previously affected the still bound electron. Consequently, the starting conditions of the electron are $\mathbf{x}(t = t_0) \cong 0$ and $\mathbf{v}(t = t_0) \cong 0$. After that, the electron moves like a Newtonian particle, its motion solely determined by the external strong field, while the atomic potential around the parent ion is neglected.

Using these assumptions, the problem becomes manageable, and the success of predictions from this simple model speaks for their validity.

a Ionisation

Experiments in [6], as well as modern routine experiments, use laser pulses in the near infrared as driving fields for HHG. Single photon energies in this regime are vastly insufficient for ionisation of, e.g., argon (first ionisation energy $\sim 15.7\,\mathrm{eV} \gg \hbar\omega\,(\lambda = 800\,\mathrm{nm}) \approx 1.55\,\mathrm{eV}$), so a multi-photon process is necessary to trigger HHG. However, the already mentioned absence of a $\chi^{(n)}$-dependence of harmonic intensities suggests that ionisation probabilities remain of the same order over a large energy interval, leading to the observed plateau structure. The competing process is the production of a free electron by tunnelling through the potential barrier formed by the distorted atomic potential (see **a** in Fig. 2.2). This is, as comparison of model and experiment reveals, the important pathway in HHG.

Multi-photon absorption and tunnelling dominate atomic ionisation in different regimes. These are determined by the relative importance of the ionisation potential I_p of the particular atom over the field energy of the external driving field. The latter is characterised by the ponderomotive potential,

$$U_p = \frac{e^2 E_0^2}{4 m_e \omega^2} \tag{2.1}$$

with e and m_e representing the charge and mass of the electron, and E_0 and ω the peak field strength and angular frequency of the electric field, respectively. This quantity calculates as the mean kinetic energy a free electron can obtain in an electric field oscillating at ω. Now the above mentioned distinction separates the following cases. When, on the time scale of an electron orbiting its nucleus, the external field feels almost static (i.e., at low frequencies), for a strong given E_0, a barrier for tunnelling is formed, making tunnelling the dominant process. In this scenario, the ionisation rate is *time dependent*, with an emphasis on the high field points of the oscillation period. For high frequencies on the other hand, when electronic orbital velocities are too slow to follow the temporal change of the electric field, multi-photon ionisation dominates. Multi-photon ionisation rates scale with the period-averaged intensity of the electric field, and have, as a result, *no time dependence* within the optical period of many-cycle pulses.

A simple model of balancing the partaking fields led Keldysh [42] in 1965 to introduce a parameter γ,

$$\gamma = \sqrt{\frac{I_p}{2U_p}}, \qquad (2.2)$$

to provide a rule of thumb for this distinction. If $\gamma \ll 1$, tunnelling is more influential than multi-photon ionisation, and *vice versa* for $\gamma \gg 1$. It is interesting to note that for the intensities and atoms used in the most common HHG experiments, γ is actually in the region of 1, without incriminating the validity of the described model. This only denotes that the whole effect is actually an hybrid of all of the possible responses of the quantum-mechanical system to the given electric field, whereas the definite 'multi-photon *or* tunnel ionisation' treatment is just one, however astonishingly successful, approximation. Time-dependent tunnelling ionisation rates of hydrogen have been modelled by Keldysh in the above mentioned treatment as early as the 1960's. The extension to other atoms was formulated shortly after, but is more famously quoted and used in [4] from Ammosov, Delone and Krainov's treatment from 1986 [43].

b Propagation

The second step starts with the assumption of an electron resting at the origin $x_0 = 0$. This actually denotes the location of the parent ion, but, as stated, SFA now neglects the influence of its Coulomb potential. We restrict ourselves to monochromatic, linearly polarised light fields, with a vanishing gradient in the relevant spatial region,

$$\mathbf{E}\left(\mathbf{x}, t\right) \equiv E\left(t\right) = E_0 \cos\left(\omega t\right). \qquad (2.3)$$

Also we detach the time scale from the temporal behaviour of the electric field E by defining $t_0 = 0$ and accounting for different phases of the light field at the time-of-birth by adding a phase $\phi \equiv \phi\left(t_0\right)$ to the argument of the cosine in Eq. (2.3).

With these assumptions, one can calculate the trajectory of the electron by integration of

$$m\ddot{x}\left(t\right) = -eE\left(t\right) \qquad (2.4)$$

as

$$v\left(t\right) = -\frac{e}{m} \int_0^t \mathrm{d}t' E_0 \cos\left(\omega t' + \phi\right), \qquad (2.5)$$

$$x\left(t\right) = \int_0^t \mathrm{d}t'' v\left(t''\right). \qquad (2.6)$$

The necessary condition for recombination is the return to the location of the ion, such that $x\,(t > t_0) = 0$. It is already clear from Eq. (2.5) that a non-zero phase ϕ will break the temporal periodic symmetry in the trajectories

$$x\,(t) = \frac{eE_0}{m\omega^2}\left[\cos\left(\omega t + \phi\right) - \cos\left(\phi\right)\right] + \sin\phi \cdot t \qquad (2.7)$$

such that, for a large number of trajectories, the return condition cannot be fulfilled. This means that, except for a select few values of ϕ, the once freed electron cannot recombine with the ion.

c Recombination

Keeping the simplified picture of an electron as a charged particle with a deterministic trajectory that *will* lead to recombination with its parent ion at the first non-trivial zero crossing of Eq. (2.7), the energy that is converted into an High-Harmonic photon is then given by

$$\hbar\omega = E_{\text{kin}} + I_p. \qquad (2.8)$$

Here, I_p is the ionisation potential, or the energy offset of the re-occupied atomic ground state in relation to the vacuum. E_{kin} is related to the square of the electron's velocity $v\,(t_{\text{rec}})$ at recombination. The time dependent ionisation rate—assuming that tunnelling is the dominant ionisation scenario—from step **a** and the resulting trajectories from step **b** lead to the observed spectral structure of recombination energies and resulting photons, such as the plateau and the cut-off. The cut-off is readily obtained by calculating the maximum attainable energy, while fulfilling the return condition, to be

$$\hbar\omega_{\text{cut-off}} \approx 3.17\,U_p + I_p. \qquad (2.9)$$

The spectral feature that remains to be explained is the observed comb structure and the spectral periodicity that leads to specifically *harmonic* high frequency components in the first place. This feature also stretches the deterministic single atom model presented above to its limits, in that one would imagine a single electron following exactly one of the possible trajectories. The more accurate, quantum-mechanical description of an electronic wave packet and its precise temporal evolution in the given external driving field, which shall be discussed in more detail in the following section, was already recognised and alluded to by Corkum. Out of experimental necessity, the first High-Harmonic observations were made using intense pulsed lasers where a large number of optical cycles were strong enough to lead to significant ionisation rates. Let us assume many equally intense cycles affecting a single electronic wave function in an initial atomic ground state.

The object that is responsible for the emission of light from this system is the time-dependent dipole operator, or its expectation value $\langle \psi \mid e \cdot r \mid \psi \rangle$ [4], where $\psi = \psi_g + \psi_c$ is composed of the wave function in the ground state and in the continuum, respectively. With some simplifications, the wave function at the origin, placed at the position of the ionic core, can be expanded to contain harmonic parts. This is due to the periodicity of the continuum wave function being determined solely by the driving field, when the SFA is applied. This picture can be expressed to incorporate the inversion symmetric three-step model *for each individual half cycle* into one self-contained representation. Radiative emission from this periodic dipole would accordingly be expected to lead to the enrichment of the driving field with discrete harmonics. The $T/2$ temporal periodicity, where T is the duration of the full optical cycle, is connected via the Fourier transform to $2f$ spectral periodicity ($f = 1/T$). So, starting with the fundamental, or '1st harmonic', the High-Harmonic spectrum will contain only the *odd integer* multiples of f. This explained the original observations to a nicety.

Owing to the simplifications of the model, this is only exactly valid if inversion symmetry for the half cycles is complete. Many experiments have overcome the odd-harmonic restriction by breaking the symmetry, either by replacing target atoms with non-centro-symmetric molecules or distorting the driving field with some low even-harmonic contributions to produce half cycles with distinguishable features [44]. Also, the restriction of HHG down to a very few or even single optical cycles via several schemes forms the basis of single attosecond pulse generation [2, 8].

Basic concepts of the quantum-mechanical model

Still concentrating on the response of a single atom to an intense laser field, we now want to outline the treatment of this situation on quantum-mechanical terms. This turned out to run more or less on the same lines as the semi-classical three-step model explained above, and was first introduced by Lewenstein *et al.* in its currently used form [45,46]. One defining aspect of HHG that runs the risk of being overlooked in the classical picture is that the full process of *all* three steps needs to be interconnected and, as such, completely coherent. The solution restated in the following section shows this at first sight.

The main benchmark of a quantum-mechanical treatment was to reproduce the experimentally ever more strongly validated finding of the cut-off law of Eq. (2.9). At the same time, one did not want to completely abandon the descriptive three steps with their respective approximations for a more general, bottom-up approach. This would necessarily be given by a full solution to the time-dependent Schrödinger equation (TDSE) for the problem[1],

$$i\partial_t \left| \psi\left(\mathbf{x}, t\right) \right\rangle = \left[-\frac{1}{2}\nabla^2 + V\left(\mathbf{x}\right) - x \cdot E_0 \cos\left(\omega t\right) \right] \left| \psi\left(\mathbf{x}, t\right) \right\rangle. \tag{2.10}$$

In the potential term of the bracketed Hamiltonian, $V\left(\mathbf{x}\right)$ stands for the pure atomic potential for the given species, while the second term represents the dipole interaction, in natural units, of the electronic wave function with the electric field, which is again assumed linearly polarised in the x-direction. Again, from the get-go, several simplifying and justified assumptions are made, namely neglecting all bound states inside the atom except the ground state $\left|0\right\rangle$, assumed a spherically symmetric state. This goes hand in hand with concentrating on the tunnelling limit of the description of ionisation—approximated by $2U_p \geq I_p$ or $\gamma < 1$, as stated above. It also means that the model is especially valid for the actual higher harmonics, i.e., those for which $\hbar\omega \gg I_p$.

As had been sketched by Corkum earlier, the wave function splits into a ground state part and all possible continuum states, labelled by their kinetic momentum $\left|\mathbf{v}\right\rangle$, and being solutions to the Schrödinger equation without the influence of the atomic potential. It can be written as [45]

$$\left| \psi\left(t\right) \right\rangle = e^{iI_p \cdot t}\left(a\left(t\right)\left|0\right\rangle + \int \mathrm{d}^3\mathbf{v}\, b\left(\mathbf{v}, t\right)\left|\mathbf{v}\right\rangle \right), \tag{2.11}$$

where a and b are the respective amplitudes of the ground and continuum states. With negligible depletion of the ground state ($a\left(t\right) \cong 1$), we nearly arrive at the solution for a free electron in an oscillating field—just as expected for the continuous part due to SFA—but with the additional influence of both the transition from the ground state to a continuum state, and vice versa,

$$\mathbf{d}\left(\mathbf{v}\right) = \left\langle \mathbf{v} \mid \mathbf{x} \mid 0 \right\rangle \text{ and c.c.}, \tag{2.12}$$

or, in this case, their respective x-components d_x and d_x^*.

With these solutions to Eq. (2.11), one can then calculate the dipole moment as defined above, although in different notation [45],

$$\begin{aligned} x\left(t\right) &= \left\langle \psi\left(t\right) \mid x \mid \psi\left(t\right) \right\rangle \\ &= i\int_0^t \mathrm{d}t' \int \mathrm{d}^3\mathbf{p}\, E_0 \cos\left(\omega t'\right) \times \\ &\quad \times d_x\left(\mathbf{p} - \mathbf{A}\left(t'\right)\right) \times \exp\left[-iS\left(\mathbf{p}, t, t'\right)\right] \times d_x^*\left(\mathbf{p} - \mathbf{A}\left(t\right)\right) + \text{c.c.}. \tag{2.13} \end{aligned}$$

---

[1]In this short section, keeping in line with the notation of the original work, theoretical in nature, we shall use the common convention of natural units, making $e = \hbar = m_e \equiv 1$.

Here we have introduced the canonical momentum in its typical form, using the vector potential $\mathbf{A}(t)$ associated with the laser field, as

$$\mathbf{p} = \mathbf{v} + \mathbf{A}(t). \tag{2.14}$$

S in Eq. (2.13) is the quasi-classical action and is given by

$$S(\mathbf{p}, t, t') = \int_{t'}^{t} dt'' \left(\frac{[\mathbf{p} - \mathbf{A}(t'')]^2}{2} + I_p \right). \tag{2.15}$$

Apart from the offset defined by I_p, this action phase term is identical to that of an electron moving in free space. So, in conclusion, Eq. (2.13) incorporates all aspects of the intuitive three-step model:

- the transition probability amplitude from the ground state to a continuum state at time t', overlapped by the driving electric field amplitude at the same time,

- the phase gained during free space propagation between times t' and t, and

- the recombination probability amplitude going back into the same ground state at time t.

Following the argumentation from [45], because full numerical solution of the TDSE at that time was still cumbersome, an effort was made to bridge back to an almost classical picture, by considering only stationary points in the quasi-classical action. This reads along the same lines as in classical mechanics, where the variation over all possible trajectories of an object under the influence of external forces leads to its *actual* trajectory by minimising the action. In this picture, it comes down to the exact same thing, as it can be interpreted by the trajectories of a classical electron contributing to a certain return energy, present in the emitted spectrum. Mathematically, the assumption can be expressed as

$$\nabla_{\mathbf{p}} S(\mathbf{p}, t, t') = 0, \tag{2.16}$$

and can be associated with the return condition of the classical model. Solving of the resulting set of equations led to the realisation that two trajectories—one with a short time $\tau = t - t'$ that the electron spends in the continuum, and one with a longer time—contribute significantly to the stationary points, or one specific recombination energy. Under certain conditions, there exist even more solutions that fulfil Eq. (2.16), however these compete more heavily with quantum diffusion of the electronic wave packets and, thus, have lower recombination probability amplitudes [46].

Note that the definition of $x(t)$ in Eq. (2.13) is, in natural units, equivalent to the temporal behaviour of the dipole moment $\mu(t)$. From its Fourier transform, the spectral dipole moment is obtained, which, assuming periodicity, can be formed discretely [4],

$$\mu_M \sim \int_0^{2\pi} \mathrm{d}t\, \mu(t) \exp(\mathrm{i}Mt), \qquad (2.17)$$

where M denotes the energy of the Mth harmonic. Using the $\mu(t)$ from above, this should, up to some factors, already be enough to provide a qualitative picture of the single atom harmonic response as it was sketched in Fig. 2.3, within certain limits, given the large number of approximations. Especially for the highest harmonics this is a good description, and as such is qualified to obtain the strength of their relative contributions to the dipole moment, as well as the cut-off law. A lengthy calculation leads to very good agreement with Eq. (2.9), up to a small functional dependence of the term containing the ionisation potential on U_p—a great success of this quantum-mechanical treatment.

Phase and propagation effects

We have seen that the overall spectral structure of HHG is obtained by describing the interaction of a single atom with an intense short laser pulse. In order to benefit from frequency-converted light on macroscopic scales, nonlinear optics is subject to certain parameters which ensure that as much of the converted light as possible is present in the far-field of a beam. This means ensuring that the phase velocity of the generated light is matched to that of the fundamental such that it adds up constructively at every point along the propagation direction. In this ideal case, the coherent superposition of a large number of the single particle events described by the three-step model leads to a macroscopic build-up of an actual High-Harmonic beam.

The formal approach to this is derived by writing down the evolution of the field amplitudes in propagation, starting from Maxwell's equations inside a material, and describing the induced polarisation as a source of fields oscillating at new frequencies. In the relatively simple case of three participating frequencies, for instance in sum-frequency generation, it turns out that the field amplitude of the signal is modulated by a term [47]

$$\Delta\mathbf{k} = \mathbf{k}_1 + \mathbf{k}_2 - \mathbf{k}_3, \qquad (2.18)$$

where the $\mathbf{k}_i$ are the wave vectors of the two fundamentals and the sum, respectively. In general, outside of a vacuum, $\Delta\mathbf{k}$—called the phase or wave vector mismatch—will be non-zero, causing the amplitude of the generated sum frequency to periodically change between zero and a comparatively small value. However, if the condition

$$\Delta \mathbf{k} \approx 0 \tag{2.19}$$

is met, the signal grows quadratically with propagation distance inside the nonlinear medium. In this case, *phase matching* is fulfilled.

Equation (2.19) can also be read as the law of conservation of momentum for the photons participating in a nonlinear optical process. In the example of Second Harmonic Generation (SHG) in a bulk crystal, the matching of fundamental and signal is commonly obtained by using a material with birefringent properties. Most of the time, a certain incident angle with respect to the optical axis can be found where, for the fundamental and second harmonic at either ordinary or extraordinary polarisation, the refractive indices for both waves are the same.

In the case of HHG as described above, the case is a bit more complicated. Formally, the phase mismatch of the Mth harmonic is given by

$$\Delta \mathbf{k} = M \cdot \mathbf{k} \left(\omega_{\text{fund.}} \right) - \mathbf{k} \left(M \cdot \omega_{\text{fund.}} \right). \tag{2.20}$$

We now restrict ourselves to conditions as they are most commonly realised in typical collinear HHG schemes in our laboratories. This comes down to using a noble gas as interaction medium, and focussing an ultrashort pulse of 800 nm central wavelength to intensities in the range of several 10^{14} W/cm^2 into the proximity of a small tube containing the gas. The dispersive terms that have to be included into the phase matching calculations are as follows.

- The equivalent to the dispersive terms in the bulk crystal example of SHG above is *neutral dispersion* within the gas. It is dependent upon the atomic number density, which in experimental conditions is controlled via the gas pressure in the interaction region. In the near infrared, where our driving laser is operating, the refractive index is usually higher than in the far UV where the harmonics are created, so the contribution to Eq. (2.20) will be positive,

$$\Delta k_{\text{disp.}} > 0. \tag{2.21}$$

- At our given laser intensities, which are a prerequisite for HHG, a large number of free electrons are created in the medium. The majority of these does not fulfil the recombination condition, so the medium becomes a partially ionised plasma during the interaction. *Free electrons* will always lead to a negative term in the phase mismatch,

$$\Delta k_{\text{plasma}} < 0. \tag{2.22}$$

- The use of focussing optics is indispensable for obtaining HHG intensities. Paraxial Gaussian optics consistently describe the amplitude and phase of the electric field when going through a focus. The curvature of the phase front, or the variance of the local wave vectors, must switch from converging to plane to diverging to meet this description. In propagation direction, along the optical axis, this leads to a spatial phase term that a focussed beam obtains with respect to an hypothetical plane wave travelling the same distance,

$$\Phi\left(z\right) = -\arctan\left(\frac{\lambda z}{\pi w_0^2}\right), \tag{2.23}$$

called the *Guoy phase*. Here, the focus is located at $z = 0$, and w_0 is the Gaussian waist radius, which, in practise, is determined by the focussing conditions such as lens power and collimated beam radius. The distance Δz between the focus and the region where the main interaction takes place—usually the region with the highest gas density—determines what phase value of Eq. (2.23) will enter the phase matching condition. It can be shown that this geometric term is

$$\Delta k_{\text{geom.}} > 0 \tag{2.24}$$

under the condition from Eq. (2.20) and certain assumptions regarding the macroscopic process.

Figure 2.4 depicts the typical geometric conditions for HHG in all experiments in this work, while also showing the handle for all relevant phase matching parameters in such a setup. The interaction happens inside of a tube that is supplied with a noble gas from one side. Laser-drilled holes allow the fundamental laser to be focussed near the centre of the gas-filled tube. The overall intensity is given by the laser power and the refractive power of the lens, including small variations due to aberrations from non-perfect focussing conditions. The intensity in turn determines the fraction of ionised gas and thus the influence of $\Delta \mathbf{k}_{\text{plasma}}$. This is a rather inert parameter, as for most experiments the highest possible peak intensity is desired, for obtaining the highest cut-off energies. The neutral dispersion term $\Delta \mathbf{k}_{\text{disp.}}$ can be easily accessed by changing the gas pressure p_{gas} and density, while it also needs to be balanced such that enough target material is provided for HHG in the first place.

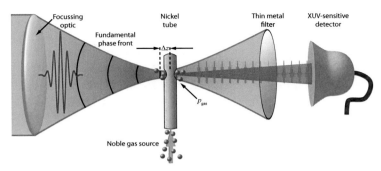

Figure 2.4: Sketch of the experimental principle for High-Harmonic Generation in a gas tube, as is relevant for all parts of this work. Focussed ultrashort, near-infrared laser pulses come from the left and interact with a noble gas that is fed into a nickel tube. The holes in the tube walls are drilled before each experiment using the laser itself in the same setup. Typically, their diameter is between 50 and 100 μm. The fundamental is filtered using thin (200–300 nm) metallic foils of a material matching the desired transmission of XUV light. The accessible phase matching parameters are the backing pressure p_{gas} of the gas feeding and the relative position Δz of gas and focus.

The most interesting tweaking parameter, however, is the position of the centre of the gas distribution relative to the position of the focus Δz. Of course, if the exact focus location does not overlap with the bulk of the target material this will lead to lower intensity in the interaction region, which influences $\Delta \mathbf{k}_{\mathrm{plasma}}$. On the other hand, it also influences $\Delta \mathbf{k}_{\mathrm{geom.}}$, which is known to have an effect on the selection of quantum paths that are responsible for the observed harmonic signal [48–50]. Harmonics from different quantum paths have distinct features. In most of our cases the short trajectories are preferred, which corresponds to a positive Δz, or focus in front of the densest gas region.

Compared to phase matching in low order nonlinear processes, as for example introduced here in Eq. (2.18), the actual conditions for gas harmonics are much harder to model. For one, a dynamic gas plume inside a vacuum chamber has a shifting, specific density profile, rather than the clear-cut coordinates of an SHG crystal. Within this density map, the phase entering $\Delta \mathbf{k}_{\mathrm{geom.}}$ can vary strongly, even switching from inside to outside of the Rayleigh range of the laser focus—typically of the order of a few millimetres under our conditions. Add to this the extreme discrepancy between the fundamental and harmonic wave vectors, phase matching can then usually be achieved more or less accurately only for a few of the High-Harmonics. Folded by re-absorption in the gas and filter transmission curves (see Fig. 2.4), a typical spectrum of High-Harmonics can be seen in Fig. 2.3 **b** for our experimental conditions.

This brief overview of the sprawling topic of High-Harmonic Generation is merely scratching the surface of all that can be said about this still relatively young field. However, this should be sufficient to allow for a complete understanding of the further parts of this work. Before returning to some of the points made here in section 2.4 dealing with HHG in the near fields of plasmonic nano-antennas, we shall first give an introduction to plasmonic effects, especially in nanoscopic particles, in the following two sections.

2.2 Surface Plasmons and Field-Enhancement

One of the main properties driving research in plasmonics is the ability to concentrate and enhance electromagnetic fields, with degrees of freedom in shaping these fields that transcend free-space electrodynamics [51, 52]. Technology benefits from this research in such diverse fields as biosensing [53], near-field microscopy/spectroscopy [54] or efficient light coupling [55]. We shall now give a brief introduction to the physical mechanisms that form the groundwork of this field.

Prerequisites for plasmonic effects

We start our considerations by looking at the bulk material properties that make plasmonics possible in the first place. The common defining properties of all plasmonic effects arise from how collective quasi-free electrons inside a metal respond to an external electromagnetic excitation. A material is considered a metal or metal-like if it has atomic bands that cross the Fermi energy. This means that free electronic states exist in the immediate energetic neighbourhood of occupied bound states. No threshold energy is required to excite electrons to these free states, which means that the metal is conducting. The well-known Drude model describes these electrons as particles of an electron 'gas' that moves under the influence of potentials and external exciting forces. Their collective net motion is associated with electric current, and properties such as the conductivity of the metal are governed by the material-specific potentials, which in turn result from the band structures in more refined models. Note that this model is exceptionally good for the noble metals such as gold and silver. Their inner, fully occupied bands lead to an effective screening of the ionic lattice, such that it is justified to describe it as a weak, uniform background to the motion of (quasi-) free electrons.

Another connected property of a metal is its response to specifically light-like electromagnetic fields. In a nutshell, the Drude model comes down to writing the Newton equation for a single electron driven by an oscillating electric field[2] [56]. The dissipative term then contains a parameter γ which incorporates the net influence of the surrounding material. The global polarisation is obtained by multiplying the single-electron displacement with the elementary charge and the average carrier density.

As a consequence of the immediate coupling of the free electrons to an electromagnetic excitation, metals are, in everyday experience, non-transparent. This is valid for a large excitation frequency range, up until a global resonance, depending on the chemical properties of the bulk metal. This is called the plasma frequency [57]

$$\omega_p = \sqrt{\frac{N_e e^2}{m_e \varepsilon_0}},$$ (2.25)

where e, m_e and ε_0 are the charge and mass of the electron and the permittivity of free space, and the material specific factor N_e is the electron density. From the treatment outlined in the preceding paragraph, this quantity is identified in the dielectric function,

$$\varepsilon_m(\omega) = 1 - \frac{\omega_p^2}{\omega^2 + i\gamma\omega}.$$ (2.26)

By approximating Eq. (2.26) to small dissipative losses $\gamma \to 0$ one can see that, at frequencies above ω_p, the dielectric function is real, and $0 < \varepsilon_m < 1$. In terms of optics this means that the metal becomes transparent. One can imagine this as the point where the electrons have become too inert to follow the electromagnetic field that drives their motion. On the other hand, there is no propagating solution for light with frequencies $\omega < \omega_p$ hitting a metallic surface, which is described by a negative real part of the dielectric function

$$\mathrm{Re}\,\varepsilon_m < 0.$$ (2.27)

For all materials of interest here, we shall also require that dissipative losses in the material are small, which means for the imaginary part of the dielectric function,

$$\mathrm{Im}\,\varepsilon_m \ll -\mathrm{Re}\,\varepsilon_m.$$ (2.28)

Again, this is a common property of noble metals, and indeed the restriction to noble metals as examples of 'good' plasmonic materials is practical for all further purposes in this work [58]. For justification, see the numbers presented in Table 2.1.

The immediate coupling of electromagnetic waves to a plasmonic material and its reciprocal action back on the net field overcomes limitations hitherto attributed to freely

---

[2]The magnetic part of the field only begins to play a role when electron velocities approach the relativistic regime. This will not be an issue for any part of this work.

	$\operatorname{Re}\varepsilon_m$	$\operatorname{Im}\varepsilon_m$
Gold @ 1.55 eV	-24.1	1.51
Silver @ 1.55 eV	-31.0	0.41

Table 2.1: Real and imaginary parts of the dielectric function at photon energies around our typical 800 nm sources of two of the most important plasmonic (noble) metals, gold and silver [59].

propagating waves. As the spatial distribution of electrons is limited by the Pauli principle on a much smaller spatial scale than the limit to focal spot sizes on account of the Abbe criterion, concentration of electromagnetic energy in the form of electro-mechanical energy on far sub-wavelength scales is possible. The result then only depends on the geometry of the participating plasmonic objects, which is indeed the main point of diversity and versatility of this field.

Localised plasmons—Mie theory

The simplest example of this interaction was already fully analytically treated by Mie in the very early twentieth century in the theory that was subsequently named after him [60]. The model system consists of a homogeneous metallic sphere inside an electromagnetic field (Fig. 2.5 **a**). The previously mentioned negative real part of the sphere's dielectric

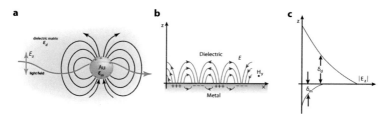

Figure 2.5: Illustration of basic plasmonic effects. **a**, Sketch of the prerequisites for describing a localised (particle) plasmon according to Mie's theory. The charges inside a metallic (gold) sphere of dimensions $< \lambda$ are excited to oscillate by an external electric field. In lowest order, for $2R \ll \lambda$, the phase of the electrons is uniform over the entire volume. The resulting near field is dipolar in nature (after [61]). **b**, **c**, Schematic of a Surface Plasmon. On the metal surface, charges are periodically separated, inducing fields in both the dielectric and the metal. These are evanescent and, consequently, confined in the direction perpendicular to the interface. The decay and degree of confinement is near the diffraction limit in the dielectric half space, but determined by the skin depth on the metallic side (from [30]).

function for $\omega < \omega_p$ leads to a non-propagating, decaying field solution inside the metal. Inside bulky material, this results in a conducting 'skin' and a field free core within the metallic object. The spheres in Mie's theory however should be of radii R less than the characteristic decay length called the skin depth [58]

$$l_s = \frac{\lambda}{2\pi} \left[\mathrm{Re} \left(\frac{-\varepsilon_m^2}{\varepsilon_m + \varepsilon_d} \right)^{1/2} \right]^{-1}, \tag{2.29}$$

where ε_d is the dielectric constant of the material surrounding the metal. In this case, the external field can penetrate the spherical particle, and in addition will be at approximately the same phase over the extent of the sphere if the wavelength is also assumed as $\lambda \gg R$. The theory now consists of solving Maxwell's equations for this exact geometric problem. Graphically, the collectively oscillating quasi-free electrons act in lowest significant order as a dipole, adding to the total field outside the particle. This dipolar, particle-like response has a resonance for

$$\mathrm{Re}\, \varepsilon_m = -2\varepsilon_d, \tag{2.30}$$

provided that the constraint of Eq. (2.28) holds [62]. The observations from this model system are a concentration of energy which is, as mentioned, governed by the size of the particle as opposed to the focusability of light at a given wavelength. In fact, the field strength in the immediate vicinity of the particle scales favourably for smaller particles, which will be of importance later in this section. Also, as the resonance is rather sharply defined by chemical properties of the material as well as particle size, the colourful scattering of colloidal solutions of these nano-scale metallic spheres, which motivated Mie's treatment of this subject, is observed.

From virtually point-like spheres to larger ones, Mie covered the whole range of multipolar expansion, where higher orders gain importance as the electron distribution takes on more complex shapes, and relative phases of electron oscillation in different spatial regions begin to matter. Furthermore, one can introduce asymmetry parameters in order to cover elliptical particles and their degeneracies such as disc- and cigar-shaped objects, which is still analytically feasible [57]. Even more complex geometries benefit only conceptually from Mie's considerations. The underlying phenomenon, however, can occur fundamentally, independent of these localised, particle-like examples, wherever an electromagnetic field is present at the interface between a dielectric and a metal. The combined fields in both media are called *Surface Plasmon Polaritons*, or shorter *Surface Plasmons*.

Surface plasmons

A Surface Plasmon (SP) (Fig. 2.5 **b**) is a mode that can propagate freely parallel to a metal-dielectric interface, but has an imaginary wave vector component perpendicular to that, in the dielectric as well as in the metal half-space [63]. While the dielectric function inside the metal always restricts the solution of Maxwell's equations to this type, excitation of a so-called evanescent mode in a dielectric is achieved, for example, by total internal reflection from an optically thicker medium. Under these conditions, the electromagnetic field is confined to a very narrow region around the interface, while being allowed to occupy all states of wave vectors along the interface that fulfil the special dispersion relation [30]

$$k_{\mathrm{SP}} = \frac{\omega}{c} \sqrt{\frac{\varepsilon_d \varepsilon_m}{\varepsilon_d + \varepsilon_m}}. \tag{2.31}$$

This usually results in breaking of the diffraction limit for the used wavelength in the perpendicular dimension (Fig. 2.5 **c**). The constraints on the wave vectors, especially on the dielectric side, require special arrangements to excite propagating SP modes. Already from Eq. (2.31) it can be seen that, for typical values of the dielectric functions[3], the magnitude of the SP wave vector is greater than the free space value. One can fulfil wave vector matching, for example in special configurations, by using total internal reflection as mentioned earlier, coupling light through a glass prism into a thin air (or lower refractive index) gap with an adjacent metal layer, or similarly *through* a thin metal layer with air or a dielectric beyond that [56]. Additional periodic structuring on length scales of the inverse k_{SP} can enhance the propagation length determined mainly by the dissipation within the metal. More complex response, similar to photonic materials, can be engineered by periodic patterning in both transverse dimensions [31].

Field enhancement

We have already seen that the coupling of electromagnetic to electro-mechanic energy by nature overcomes the free space diffraction limit of optics. The feature size of field gradients is governed by the dimensions of the coupling object. A result from Mie's theory in lowest order, as it was previously described to be equivalent to dipole emission, is the near field term (for $k \cdot r \ll 1$) of the electric field [64],

$$\mathbf{E}_{\mathrm{near}} = \frac{3\mathbf{n}\,(\mathbf{n} \cdot \mathbf{p}) - \mathbf{p}}{4\pi\varepsilon_0\varepsilon_d}\,\frac{1}{r^3}, \tag{2.32}$$

---

[3]That is, $\varepsilon_d > 0$ and $\varepsilon_m < 0$ for $\gamma \to 0$, $\omega < \omega_p$, cf. Eq. (2.26).

where **n** is the unit vector in the direction of the respective spatial coordinate and $\mathbf{p} = q \cdot \mathbf{x}$ the dipole moment. For very small spheres, this divergence is significant, and consequently the field in the immediate vicinity of the particle ($r \approx 0$) is strongly enhanced.

A classical pendant is evident in everyday experience. In electrostatics, spatially strongly variant metallic objects, such as tips, result in high potential gradients and, thus, strong fields in their surroundings. This is known as the 'lightning rod' effect. Although the scale and material assumptions for this intuitive picture—negligible skin depth and perfect conductivity—have been shown not to apply in the nanoscopic regime, it can be readily used as an intuitive description here, while the more rigorous approaches mentioned later incorporate all of these specifications. Large electron concentrations, crammed into tiny spatial features and, moreover, resonantly excited, lead to time-dependent high field enhancement, which is an immediate basis for the enormous number of applications mentioned in the introduction. We shall go into the specifics of these field shapes in the next section.

In most real-world examples the bulk effects as well as the exact surface behaviour of the modes inside a metallic particle are determinant, each influencing the other. Thus, it is practically impossible to give analytic results for the response of metallic particles of arbitrary shapes and sizes below or similar to the exciting wavelength. Simple breaking of the full spherical or cylindrical symmetry of Mie's theory was already hinted at. Another way of comprehending the spectral response of basic yet non-trivial geometries is by breaking down the problem into manageable components and viewing the result as a hybridisation of each of these components [57]. Closer to the numerical side, one can also discretise the entire geometry and calculate the overall response from single, correlated Mie-like dipolar elements, a procedure that is known as *Discrete Dipole Approximation* (DDA) [65]. With today's computing powers it is more common to use the most basic approach, namely the full, time-step-wise solution of Maxwell's equations in an arbitrarily fine grid building up the full geometry of the problem. This method is known as *Finite-Difference Time-Domain* or FDTD [66].

Harnessing the potential, especially by taking advantage of large enhancement factors, the tailoring of electromagnetic near fields has become a driving force for research. As mentioned earlier, the ability to simulate arbitrary geometries on reasonable time scales, as well as the vastly improved nano-structuring techniques, have helped make this field as diverse as it is today. Commonly, the similarities to metallic objects that react resonantly to long wavelength electromagnetic radiation have led to the naming of these tiny tailored objects as *Nanoantennas*. While this comparison is intuitive, it is a bit of an understatement for two reasons, firstly because the dimensions on length scales of the skin depth (Eq. (2.29)) make the resonance behaviour less predictable, and secondly because effects unknown from conventional antenna physics are introduced by the relative importance of the magnetic near fields [67]. These effects, compelling as they are, cannot be addressed here, as they are far beyond the scope of this work. We shall, however, treat certain types of antennas, especially the aforementioned bow-ties, more closely in the following section.

2.3 Metallic Nanoantennas

From radio frequency to optical antennas

An antenna, or aerial, is an object that is employed when increased coupling is desired. Traditionally, this coupling is supposed to happen between a localised electric current within a source or receiver and a radiation field. A radio can pick up and amplify signals within its internal circuits, but it is only effective when these are connected to an antenna. In the sophisticated varieties, the functionality as an optimal transducer between free radiation and localised energy dictates the design process [52].

In long range radiocommunication, distant transmitters, even from directionalised aerials, deliver only small radiative powers at the location of the receiver. In turn, a weak signal must be coupled effectively into the free field for broadcast. The simplified version of this reciprocal model is given by two elementary dipole moments coupling via an electromagnetic field such that the field produced by the emitting dipole induces a reactive dipole moment at the coordinate of the receiver. In the case of full isotropy, the power density at a distance r from the transmitter decays with the size of the surface of the pervaded sphere as r^{-2}. An antenna can, due to its geometry, improve coupling efficiency with directivity, but it is also limited in that it is always prone to dissipative losses. These two things combined characterise the *gain G* or figure of merit of an antenna [52]

$$G = \epsilon_{\text{rad}} \cdot D, \tag{2.33}$$

where

$$\epsilon_{\text{rad}} = \frac{P_{\text{rad}}}{P_{\text{rad}} + P_{\text{loss}}} \tag{2.34}$$

is the efficiency dictated by dissipation, and

$$D\left(\theta, \phi\right) = \frac{4\pi}{P_{\text{rad}}} p\left(\theta, \phi\right) \tag{2.35}$$

is the directivity, multiplying the total radiated power by an angular density characteristic. Later on, we shall add to these features the near field enhancing properties of certain geometries introduced before.

The rule of thumb for the length of classical aerials receiving or broadcasting at a wavelength λ gives roughly $\lambda/2$, while transverse spatial dimensions are usually negligible. In first approximation, this still holds true when scaling down from radio to optical frequencies. The main difference is that at some point, depending on the material, the penetration depth is similar to and even greater than the antenna size. This leads to a relative importance of the inner, evanescent modes. Electronic response is no longer limited to an infinitely thin surface, but has a non-trivial phase behaviour in relation to the external field, and can be spatially variant over the whole nanoscopic object. Transverse retardation effects must be taken into account since the aspect ratio of a nano-antenna cannot remain effectively one-dimensional either.

It has been shown that even for simple, for example cylindrical, geometries with $R \ll \lambda$, the antenna response can be approximated to fit to an effective wavelength that can be parameterised as [68]

$$\lambda_{\text{eff}} = n_1 + n_2 \left(\frac{\lambda}{\lambda_p} \right), \tag{2.36}$$

introducing geometric coefficients n_1 and n_2 with dimension of length, and λ_p the plasma wavelength corresponding to Eq. (2.25) as $2\pi c/\omega_p$. This conclusion has been reached by starting from ideal, metal wire antennas as degenerate cylinders. In sufficiently small dimensions, and with the assumptions of the Drude model, Eq. (2.36) is obtained using complicated but analytical expressions for n_1 and n_2. Notwithstanding the exact calculations that led to this result, the trend of nanoscopic antennas to be resonant at effective wavelengths that are *shorter* than the intuitively expected—for metallic antennas, factors between 2–5 had been found phenomenologically [52]—holds true for other geometries as well.

Following traditional designs, it is just a small step from the wire-, or rod-like, antennas to split dipolar antennas. Concerning radio aerials, the gap in the middle of a single antenna—impedance-matched to leave the combined $\lambda/2$ resonance of both half-antennas more or less unaltered—is necessary to feed the signal as an electric current into the antenna system. But furthermore, as was seen in the previous section, the gaps in nanoantennas, which can be manufactured to be less than 20 nm, can strongly enhance the electromagnetic near fields [51]. In practise, optical antennas are more suited to couple these fields to the far space, rather than inducing or amplifying electric currents within the gap. As such, they are a beneficial tool for exciting nonlinear optic effects in a very confined space, or detecting and amplifying weak emitters [69]. Taking that design one step further leads to the bow-tie antenna, a structure which is currently being actively researched. As this geometry is of great importance in this work, it shall now be described in more detail in the following section.

The bow-tie antenna

As mentioned in the outline introducing this part of the present work, the bow-tie antenna gets its name from the shape of the male neck ornament. It follows the above stated idea of splitting one single antenna into two parts with a separation small enough that a coupling via the real fields around the facing ends is possible[4]. As for split rod antennas of proper dimensions, the enhanced field in the gap region makes this coupling possible, while providing tightly confined, sub-wavelength high field strengths for further purposes. In fact, it has been shown that the bow-tie geometry supports greater intensity enhancement factors than comparable simple rod-type features [70], Fig. 2.6 showing the results from the cited source. This is due to the fact that the relative strength of 'hot spots' in the geometry is smaller at the blunt ends of the triangles and even more concentrated between the tips (compare the peripheral ends of the respective antennas in **a** and **b** of the aforementioned figure). Also, compared to the rod, the defining features of the tips are more acute for the bow-tie solution, while at the same time the mode volume of the entire antenna and the area of acceptance is increased.

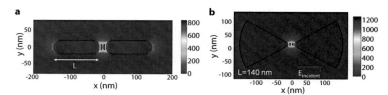

Figure 2.6: Comparison of the simulated field enhancement on **a**, a rod-type and **b**, a bow-tie nano-antenna of similar dimensions. The length *l* of a single rod and one bow-tie triangle is 130 nm and 140 nm, respectively, to match resonances. Both antennas have gap sizes of $g = 20$ nm (from [70]).

In 1996, Grober *et al.* realised and characterised a bow-tie antenna resonant to microwave radiation [71]. Most prominently, they demonstrated the antenna's ability to

---

[4]We shall, for all purposes here, concentrate on the polarisation of the exciting electric field that lies parallel to the axis connecting both triangles.

concentrate the electric field emitted from a microwave waveguide to a spot of $\lambda/10$ at a high transmission efficiency. Consequently, they proposed to transfer this result to optical frequencies by means of the ever improving lithographic techniques, provided that the aforementioned plasmonic requirements to the material were met.

We shall take a closer look at the defining features of this particular antenna type in this section. Figure 2.7 is a schematic of one single, free-standing bow-tie antenna, where all the tunable parameters are labelled. We shall then address in short the influence of each parameter.

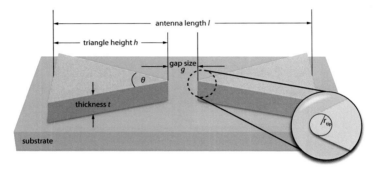

Figure 2.7: Labelled schematic of one single bow-tie antenna. Not captioned, the angle θ defines the opening of the idealised single, isosceles triangle. Due to the fabrication process, all tips and edges are rounded in actual nanoscopic samples. Most importantly, the rounding of the opposing triangle tips is characterised by the radius of curvature r_{tip}.

Antenna length l and gap size g

In first approximation, l defines the resonance behaviour of the entire antenna system, again only in response to a polarisation along the connecting axis. The approximate region where the resonance is found responds to a $\lambda/2$-rule of thumb, with significant corrections if the entire system is scaled down to near the size of the field penetration depth, as was seen in the previous section. But furthermore, the functionality of the antenna is inseparably connected to the gap size g. The interplay between l and g may depend on the chosen method of fabrication, a point that we shall address in section 3.3. Here, we consider the two options of either varying the overall antenna length by changing the size of congruent triangles while leaving g the same, or varying the distance of identical triangles, as well as the connection between both cases.

At first sight, the intuitive point holds that longer antennas are resonant to larger wavelengths. However, in the optical regime as opposed to radio frequencies, the coupling between the antenna halves is not induced by an impedance matched feed circuit, but by the strong field around the apices of the triangles. One can further separate the two cases, one where coupling is achieved, and one where the declining field strength between the apices cannot sufficiently bridge the gap. In many realisations, especially the optimal achievable scenarios in the optical regime, a transition between the two cases is observed.

In the first case, each single triangle acts as a resonant object, an asymmetrical version of rod-like structures, with a resonance that can be approximated using Eq. (2.36) or modelled by full field simulations. The resonance behaviour displays a feature that corresponds to the induced dynamic charge density oscillating parallel to the exciting field polarisation [72, 73]. In this regime, the gap only has a small influence on the position of the resonance. On the other hand, for fixed triangle sizes with a decreasing distance between the two, the system becomes—more or less strongly—coupled. Initially, opposite charge densities accumulate at the opposing features in the middle of the bow-tie. If the resulting near fields extend far enough away from the surface, their superposition counteracts, leading to a suppression of energy in the system, and thus a red shift of the plasmon resonance, in comparison to the single triangle [70, 74].

Increasing the coupling pulls the resonance away from the single antenna case, towards longer wavelengths. As might be noted from this description, there is no absolute scaling for the gap size to allow coupling, as the responsible field in the gap depends on the resonant build-up that can be supported by the individual triangle. However, there seems to be a minimum ratio of $r =$ (separating distance : triangle height), which was determined for example to be $r \sim 2$ for near-isosceles, 75 nm high gold bow-ties[5] [75]. Still, for bottom-up, arbitrary antennas, the strong interconnectivity of all of the parameters that contribute to the overall length make a full simulation indispensable.

For our purposes it should also be mentioned that for coupled bow-ties the intensity enhancement displays higher factors than the mere near field enhancement around equally sharp tips of decoupled triangles [76]. However, a limitation to the high enhancement for increasing excitation power was already observed in the cited experiment.

---

[5]In this dimensionless quantity, the separating distance is defined as the distance between the half-height centres of the triangles, so it calculates as the gap length *plus* twice the triangle half-height. Conclusively, values of r range from $1 \ldots \infty$, where 1 is the case of touching triangles.

Opening angle θ

The widening of the opening angle influences both the resonance and enhancement behaviour. First, let us consider a single triangle, and its degenerate case of the rod for $\theta \approx 0$. In the case of large opening angles, the dimension of the particle transverse to the symmetry axis gains importance. Of course, this has a strong influence on the resonance with respect to light which is polarised in this direction, but we still want to concentrate on the excitation along the tip-connecting axis of a bow-tie. Here, however, some transverse features also begin to play a role. Boundary conditions at the metal surface enforce currents along the sides enclosing θ, where the geometrical path length for carriers is $\left(\cos \theta/2 \right)^{-1}$ times that of the axial path. These currents correspond to a quadrupolar mode as they break the trivial symmetry of the axial dominant dipole oscillations. They may show up as weak, broad spectral features in the response of the bow-tie [70, 73].

When two coupled triangles in the form of a full bow-tie are considered, the aspect ratio between longitudinal and transverse features jumps by roughly a factor of two compared to the single triangles. It may be inferred that in this case, the mode which is dipolar in nature gains relevance, as is indeed observed in simulations [73].

Concerning intensity enhancement, it is shown that for smaller angles θ on the triangles the achieved enhancement factors of a typical bow-tie become larger, even by up to a few times, for $\theta \approx 0$ (linear 'rod' antenna) [72]. We have, however, mentioned the advantages of a larger effective cross-section of the bow-tie. In fact, a non-zero opening angle is the feature which ultimately allows for the introduction of smaller tip sizes without falling back to ultra-thin, virtually one-dimensional linear antennas.

Tip rounding radius r

From what has been stated thus far, it is implied that the rounding radius of the tip r_{tip} is the better suited parameter to increase field enhancement. This is very straightforward, as a smaller radius leads to a stronger confinement of field lines, or, in other words, an improvement of the lightning rod effect. The limitation to this parameter is mainly determined by the fabrication and surface chemistry of the material, and is desired for all of our purposes to be as small as possible.

Feature thickness t

So far we have neglected the dimension of depth or thickness in our considerations. This parameter is virtually non-existent in classical antennas since there are no field components that could couple to currents in propagation direction, and the infinitely small skin depth reduces the dimensionality in any case. However, in the regime of optical nano-antennas, voluminous charge densities have to be considered, and the hybrid nature of the involved near field defies a trivial description by a single polarisation direction.

For all practical purposes, a thickness is chosen or dictated by fabrication considerations that is smaller than the in-plane features. Valid simulations of the resonance behaviour of the antenna have to take its whole volume into account. It should only be mentioned here that explicitly thickness-dependent modes can be excited in complementary bow-tie-shaped apertures as self-standing spectral features [77, 78]. However, these Fabry-Pérot-type resonances are naturally limited to apertures, or slot-antennas.

We have now spoken about nanoscopic constructs with an optimised ability to produce greatly enhanced near fields on the one hand, and HHG as a high-field process on the other. Following the presentation of an experiment that combines the two, which was described in the introductory section to this part, a theoretical description has been proposed and simulated by a few groups. In the following section we shall outline the rather straightforward approach that those groups took towards accomplishing this task.

2.4 High-Harmonic Generation in Inhomogeneous Fields

The introduction already provided a rundown of the basic idea of nano-structure assisted High-Harmonic Generation from the first experiment, as well as successive publications by the same and other groups. Other than the earliest experiment which used bow-tie antennas [29, 41], and those relying on the same geometry [39, 40], field enhancement for the goal of HHG was explored on single, funnel-shaped devices [41, 79], and on linear, rod-type antennas [80]. This concept has inspired theoretical work towards the same goal, such as hypothetically exploring more exotic structures that require demanding production steps. This includes a two-fold symmetric extension of the bow-tie in form of a cross, or a regular ordering of bow-ties whose axes are oriented orthogonal with respect to each other [37]. The authors point out that the functionality of the basic bow-tie concept along two axes could lead to the achievement of circularly polarised XUV light. Another collaboration of authors has simulated HHG in the vicinity of densely packed but non-touching spheres, whereby sufficient field enhancement is achieved within the gaps [81]. We have summarised these works and their relevant parameters in Table 2.2. All groups used gold as plasmonic material, except the simulations in [35, 37] which were performed with silver, and the experiments in [79] which made use of silver nano-cones.

Central wavelengths in all of the works (800–830 nm) were derived from Ti:Sa sources.

	Group	Geometry	Enhancement factor	Cut-off
Experimental	Kim *et al.* [29]	36 × 15 Bow-tie array	$> 10^2$ (estimate)	47 nm (H17)
	Park *et al.* [79]	Single conic waveguide	$\sim 10^2$	18.6 nm (H43)
	Sivis *et al.* [40]	Large array of bow-ties	380 (best fit)	No clear HHG
	Pfullmann *et al.* [80]	Rod-type antennas	10^2 (estimate)	H5 (no gas)
Theoretical	Yang *et al.* [81]	Hexagonally ordered, non-touching spheres	$\lesssim 10^3$	~ 19 nm (H41)
	Husakou *et al.* [37]	Crossed rods	1600	up to H125 (?)
		Orthogonal bow-ties	~ 1000	?
	Husakou *et al.* [35]	Nano-cone	$\sim 10^3$	H105
		Single bow-tie	$\sim 10^2$	36 nm (H23)

Table 2.2: Comparison of the types and parameters of different plasmon-assisted HHG schemes, experimental and theoretical, since 2008. Harmonic numbers, if not given directly in the original reference, have been calculated using the respective used or simulated central wavelength, which in all cases is around or above 800 nm.

One should mention that the numbers for the cut-off energies rely mostly on very generous interpretations of the obtained spectra, especially for the simulated results in Table 2.2. Some spectral features are still counted as harmonics, and subsequently used for further calculations, while their normalised signal is below 10^{-6}, or even around 10^{-10} in the case of ref. [81] and the second part of ref. [37]. In fact, the same controversy comes up in the experimentally documented cases, too. References [39] and [40] critically deal with the earlier results, and some of their concerns shall be examined in the discussion of part I, also in the context of our own findings.

Another theoretical line of enquiry that was sparked by the initial experiment set out to explore, aside from specific geometries, the fundamental mechanism of the influence that inhomogeneous fields have on the well-established description of HHG. Especially the work by Ciappina *et al.* has helped advance this enquiry by rigorously inserting fields as they are produced around metallic nano-structures, or even more abstractly by introducing a finite inhomogeneity parameter into the Lewenstein model [36, 82, 83]. We want to follow their line of thought and document the basic ideas and results from these works here.

The Lewenstein model for inhomogeneous fields

The original formulation of the quantum-mechanical model describing High-Harmonic Generation was presented in short at the end of section 2.1. It is important to note the relative scales on which the relevant interaction happens. The decisive length scale is the spatial excursion of significant features of the electronic wave function, which might also be considered as the maximum distance from the core that an electron can have on its trajectory. For Lewenstein model calculations, the participating field is described as homogeneous across the region where electron dynamics take place. Now, some rather subtle modifications can result in greatly varied conditions and results for the trajectories—or quantum orbits—and, subsequently, High-Harmonic spectra.

For the typical approximations, we have already given the relevant definition of the dipole moment and its elements, and the further steps to find the trajectories via the condition for the saddle-point calculations in Eqs. (2.13)–(2.16). The small, finite inhomogeneity is now incorporated as a linear term, while the temporal dependence $E(t)$ has the common form $E_0 \cos(\omega t)$ times a pulse envelope function $f(t)$. The scalar potential for such an effective field [83]

$$E(x,t) \equiv E(t)\left[1 + \alpha x\right] \tag{2.37}$$

is given by

$$V_{\text{laser}}(x,t) = xE(x,t) \tag{2.38}$$

with a small α quantifying the inhomogeneity. Note that we have, as before, restricted ourselves to an x-polarised field and its respective potential, while the inhomogeneity also happens along that direction. The classical trajectories follow the Newton equation, but now with a non-vanishing field gradient[6] [83]

$$\ddot{x}(t) = -\nabla_x V_{\text{laser}}(x,t) = -E(x,t) - \left[\nabla_x E(x,t)\right]x \equiv -E(t)\left[1 + 2\alpha x(t)\right]. \tag{2.39}$$

---

[6]As in the corresponding part in section 2.1, natural units are used here.

When solving Eq. (2.39) to obtain the classical trajectories $x\,(t)$, typically the temporal integration of the field terms allows one to identify the role of the corresponding vector potential $A\,(t)$. Using the usual initial conditions ($x_0 = 0$, $\dot{x}_0 = 0$, as well as $E\,(t_0) = 0$), however, in the inhomogeneous case the simple solution

$$x\,(t) = \int^{t} \mathrm{d}t' A\,(t')\,, \tag{2.40}$$

together with Eq. (2.39), shows the role of the parameter α as [83]

$$A_{\mathrm{eff}}\,(t) = A\,(t) + 2\alpha \left[\int^{t} \mathrm{d}t'' A\,(t'') - \int^{t} \mathrm{d}t'' A^2\,(t'') \right]. \tag{2.41}$$

Here, unprimed t is the (classical) recombination time. Equation (2.41), which defines an effective vector potential in contrast to the trivial quantity, leads consequently to a multitude of new terms in the equations defined by the constraint of stationary action, known as the saddle-point equations. The solutions of these show an ambiguity in the possible drift momenta *for identical ionisation times*, where one solution increases and the other lowers the gained return energy and, ultimately, the obtained cut-off. The significance of this effect responds to the strength of the perturbing inhomogeneity α. Figure 2.8 shows the results from ref. [36]. While the arbitrary scaling, denoted next to the

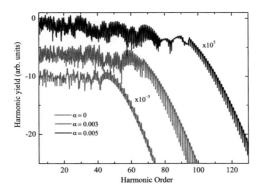

Figure 2.8: High-Harmonic spectra of an inhomogeneous field for an hydrogenic model system and monochromatic near-infrared input field with an intensity of $3 \times 10^{14}\,\mathrm{W/cm^2}$. The authors used the model described in short in this section for different inhomogeneity parameters as specified in the figure (from [36]).

respective curve, indicates that the efficiency of HHG drops significantly with the value of α, the increase in cut-off is the most remarkable feature of this study. Furthermore, although it is not clearly visible here, the breaking of the symmetry of the potential results in the additional appearance of even harmonics, better documented in ref. [83].

This has given an insight into the influence of a general inhomogeneity on the model, which has led to the most conclusive picture of the mechanisms of HHG (a similar ansatz was also presented in ref. [35], however it lacks the concise treatment of the full model). More to the point of Kim *et al.*'s original idea, the same authors have also used brute numerical solving of the time-dependent Schrödinger equation, without SFA, to simulate HHG in fields as they are obtained around bow-tie nano-structures, following the design from ref. [29]. We want to briefly describe the results here.

Numerical simulation of HHG near bow-tie antennas

For 'real' near fields around nano-structures, the inhomogeneity term has to be effective in describing the results from, e.g., an FDTD simulation. It is necessary to extend the small, linear dependence to an arbitrary shape by using a polynomial expansion, so that [83]

$$E\left(x, t\right) = E\left(t\right)\left[1 + h\left(x\right)\right] \tag{2.42}$$

with $h\left(x\right) = \sum_i h_i x^i$. Also, to circumvent the singularity, the atomic potential is typically modified as

$$V_{\text{atom}} \propto \frac{1}{\sqrt{x^2 + \xi^2}} \tag{2.43}$$

with the parameter ξ chosen to recover the proper ionisation potentials around the origin. Once the wave function $|\psi\left(t\right)\rangle$ is obtained, the full Hamiltonian is used via the correspondence principle to deliver [83]

$$a\left(t\right) = \frac{\mathrm{d}^2 \langle x\rangle}{\mathrm{d}t^2} = -\left\langle \psi\left(t\right)\left|\left[H\left(t\right), \left[H\left(t\right), x\right]\right]\right|\psi\left(t\right)\right\rangle. \tag{2.44}$$

The Fourier transform of this quantity is equivalent to the power spectrum of harmonics.

Again, in ref. [36] Ciappina *et al.* have calculated the harmonic spectrum from argon inside the gap of a bow-tie antenna similar to the ones used by Kim *et al.* They observed a significant increase in the cut-off compared to the case of an homogeneous field of the same illuminating intensity. They attribute part of that increase to the confinement of the typical HHG process by the bow-tie geometry. The full evaluation of quantum orbits contributing to a certain harmonic revealed the relative importance of very large electron excursions, especially for larger inhomogeneities, that are eventually limited by the finite space between the tips of bow-ties [82]. Furthermore, the influence of absorption of free electrons at the metallic surface itself seems to play a role in the resulting spectra [35].

This concludes the sections on the theoretical aspects of High-Harmonic Generation, in general and in the specific case of inhomogeneous fields. We have also given an introduction to the near field effects of metallic nano-particles, which are supposed to be beneficial for aiding HHG with initially lower intensity fields. A more technical but extremely important point of experiments of this type is, however, the limitation that is set upon the exposure of the typically intricate nanoscopic objects to intense laser light, which is, indeed, the main point of investigation presented in this part of the work. For this, we shall dedicate the last theoretical section to the interaction of intense electromagnetic fields with solid matter, metals in particular.

2.5 Laser-induced Damage

A very practical use of pulsed lasers especially lies in material modification and all related industrial purposes such as cutting, drilling and welding. The basis of these processes is, in a way, rooted in electrodynamics, describing the interaction of the laser field with particles of a certain material, but mainly in thermodynamics. Thus, it deals with the collective response of an ensemble of particles in a solid under the deposition of energy via laser light. The deposition is quantified by the fluence, the amount of energy per surface area, most often given as a non-SI-unit in J/cm^2.

From thermodynamic first principles, the incident laser acts as a heat source, and the material responds by heating up, which classically means an increased mean energy stored in the atomic lattice. The spread of energy within the material is governed by heat conduction, and then, for higher incident energies, where phase transitions to liquid or gas occur, the fluid and gas dynamics of melted material. When melting and evaporation is achieved, surface and bulk modification caused by rearranging the solid's structure, which can become permanent, is observed. This modification and eventual ablation—or *(Laser-induced) Damage (LID)*—is the basis for material processing by laser light, but also allows insight into fundamental solid state physics.

There is a switch of regimes when the duration of heat deposition, determined by the length of the used laser pulses, becomes much shorter than the typical time scales of heat conduction by the lattice particles. The simple thermodynamic picture of slow heating is then inadequate to account for ablation and modification processes. In this section we shall introduce the main fundamental concepts necessary to describe the deposition of ultrashort laser pulses into solid material. While being very straightforward on a theoretical level and under ideal experimental conditions, some aspects are only phenomenologically accessible.

This is especially the case when the surface structure and patterning are non-trivial, or beyond control of idealised conditions. Some specific phenomena in connection with this shall be introduced. Finally, we shall briefly describe the method of evaluation of damage threshold fluences used in the following chapter.

The Two-Temperature Model of Laser-induced Damage

For the purpose of this work, we shall constrict ourselves to the treatment of metallic solids, as do most of the fundamental studies cited here. However, many statements made by the described model also hold for ablation mechanisms of strongly excited dielectrics. In the ultrashort regime, usually multiphoton ionisation and a successive avalanching effect of produced free electrons precede the initial state from where the transport model described here begins [84, 85].

The decisive factor that separates the time scales for slow melting and ultrafast ablation is the near immediate response—compared to that of thermal diffusion by the lattice—of free electrons coupling to a radiative field. For excitation durations that are significantly below lattice diffusion constants, it is necessary to treat the energy of the electrons and the atomic lattice separately. In thermodynamic terms, two temperatures T_e and T_l are introduced, characterising the kinetic energy of free electrons in the solid and the energy of phonons or lattice oscillations, respectively [86]. The distinction is important as the lattice energy, synonymous with the macroscopic temperature of the solid, defines the points of phase transitions and, thus, possible material modifications. The threshold pulse energy E_{th} at which damage occurs is reached when the energy density deposited inside a given volume exceeds the point where a phase transition is possible. For actual ablation, a complex sequence of processes plays a role in addition to traditional melting [87]. However, for a basic understanding of the here discussed model, this simple picture should suffice.

In order to make sense of temperature as a quantity, the respective system must be thermalised. There are different processes responsible for thermalisation of both the electrons and the lattice, and the energetic coupling between them. Figure 2.9 sketches the transient states and the involved transport mechanisms for exposure of a metal surface to laser light. First, an initial highly non-equilibrium state is induced by optical transitions. From here, macroscopic dephasing of the initially coherent polarisation, as well as ballistic electron–electron (e–e) collisions, lead—after $\sim 10^{-13}$ s—to an eventual energetic (Fermi-Dirac) distribution that allows for a meaningful definition of T_e [84, 88]. Note that the exact shape of the density of states near the Fermi edge, which is, for instance, also responsible for conductive behaviour, has a strong influence on e–e collision

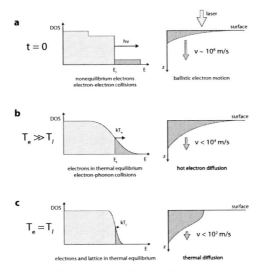

Figure 2.9: Schematic of the different stages within a metal upon exposure to intense laser light (left, the electron density of states (DOS) and right, energy distribution and diffusion). The deposition of energy happens before and ends at $t = 0$. **a**, The initial non-equilibrium state generated by photons of energy $h\nu$. **b**, Electronic equilibrium is reached at T_e, caused by e–e-collisions; subsequent thermal diffusion. **c**, The electrons heat up the lattice to T_l until $T_e = T_l$ at equilibrium (from [89]).

rates. This implies a strong material dependence of this step. Gradually, the transport of energy from the electrons to the lattice generates a global equilibrium state, characterised by the (Bose-Einstein) distribution of phonons over the Brioullin zone, which defines the eventual lattice temperature T_l. Further evolution can be seen as a thermal process, which then relaxes within some 10^{-11} s, following common heat diffusion models.

Once thermalisation of the ballistic electrons has been reached via collision processes (see Fig. 2.9 **a**), the Two-Temperature Model (TTM) describes the temporal evolution and the spatial dynamics of T_e and T_l in the following form,

$$C_e\left(T_e\right)\frac{\partial T_e}{\partial t} = \nabla\left(K\nabla T_e\right) - g\left(T_e - T_l\right) + S\left(z, t\right) \tag{2.45}$$

$$C_l\frac{\partial T_l}{\partial t} = g\left(T_e - T_l\right). \tag{2.46}$$

Here, S is the source term that describes the energy applied to the material. It is seen here to only act on the electrons directly. For real-world source terms, one usually assumes a Gaussian temporal envelope of duration τ, containing a specific (laser) pulse energy, which is reduced by bulk optical properties such as reflectance and transmittance. In more advanced modifications S can also account for the depth of deposition by including the ballistic range of the initially excited electrons [89]. The parameters C give the respective heat capacities of electrons and lattice. Other than the lattice heat capacity, C_e is significantly dependent upon the absolute electron temperature. Heat conduction in metals is dominated by the mobile electrons. Sommerfeld's model calculates K by considering combined collision rates of both e–e and electron-phonon (e–ph). Finally, the parameter g in this model is *a priori* a phenomenological coupling constant between the thermodynamic reservoirs of electrons and phonons. Corkum *et al.* published a seminal work that explored the transition from long pulse to ultrashort pulse optical damage of metals [90]. The competition of all conductive and energy exchanging processes requires careful modelling in order to keep track of each contribution.

From these considerations long pulse LID can be viewed as a special case of the TTM where excitation is maintained over time scales longer than lattice relaxation. This is a good approximation for pulses longer than nanoseconds. Most of the time, T_e and T_l are in common equilibrium, so that the intermediate states shown in Fig. 2.9 **a** and **b** lose their meaning. It is possible to describe the dynamics with one single temperature. Consequently, Eqs. (2.45)–(2.46) simplify to a single diffusion equation. In this approximation, $T_e = T_l \equiv T$ at all times, and the heat capacity is dominated by the lattice, so

$$C_l \frac{\partial T}{\partial t} - K \nabla^2 T = S\left(z, t\right). \tag{2.47}$$

Along with the maximum thermal diffusion length, E_{th} scales with the square root of the laser pulse length $\sqrt{\tau}$ in this scenario.

For sub-picosecond pulses on the other hand, the free hot electrons can diffuse deeply, long before interacting with the lattice, depending on the value of g. As a consequence, assuming that the same energy is deposited as before, but now within a few hundred femtoseconds, the lattice temperature will rise only in a narrow region downwards from the surface. At the surface itself, much higher temperatures can be reached. Thus, melting, or direct solid-vapour and plasma transitions, are confined to shallow, clearly defined volumes [87, 91]. Below a critical value τ_c, threshold fluences are independent of pulse duration. Figure 2.10 from Corkum *et al.* [90] shows the earliest findings of the transition between both regimes for copper and molybdenum.

Several consequences can be derived from these models. As long as pulse durations are well below electron-lattice interaction times, the amount of deposited energy per volume may be applied in arbitrary time frames. In other words, only the pulse energy and focussing, but not pulse shape or peak intensity are relevant for ablation. It is thus justified

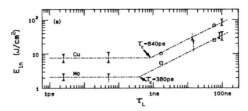

Figure 2.10: Damage threshold fluences of copper and molybdenum as a function of laser pulse duration. The switch between ultrashort and long pulse regimes is indicated for both materials by extrapolation (from [90]).

to characterise threshold values in terms of the *fluence*. In contrast, the pulse length dependence in the nanosecond regime might make it more convenient to include τ into the threshold value by calculating the respective *peak threshold intensity*. Pulsed laser sources of kHz repetition rates leave enough time for a full relaxation of the targeted material between shots. A much discussed exception of the shot count dependence of threshold fluences will be introduced shortly, as it is crucial for our experimental considerations.

Finally, as the penetration depth of excited electrons was seen to play a decisive role in ultrashort LID, an obvious constraint to this quantity is set by the physical extent of the bulk material under observation. Thin layers inhibit the free pervasion of electrons, as these encounter a boundary on the far side of the initial excitation. This results in an accumulation of energy within a smaller volume, and, for identical fluences, damage occurs earlier than in a bulk of the same material. This will be documented in the upcoming chapter, especially as our final interest lies on the sensitivity of nanoscopic objects to LID.

Some phenomenological peculiarities in Laser-induced Damage

To begin, the Two-Temperature Model is one-dimensional in the propagation direction of laser light impinging onto the material. This is usually justified since even tightly focussed spot sizes are of orders greater than the relevant depth scales of optical absorption and electron penetration. In this model, there are no polarisation anisotropies. The actually absorbed fraction of the incident light in the source term S is given by the global optical parameters of the material. However, in the physical reality, the microstructure of the crystal surface is expected to have a great influence on the local energy distribution and material reaction to the excitation. As a result, the values of the threshold fluence for the ablation of a material F_{th} can only be relevant for specific experimental conditions.

But furthermore, there are modifications that occur only around F_{th}, which might not be quantifiable, and are hardly visible by common evaluation techniques. Much like surface impurities that have been present beforehand, these *induced* structural defects change the local absorbance, and make themselves noticeable when many shots are applied to a single site. Consequently, the evaluated damage thresholds have to be interpreted by whether they were obtained in multi- (N-on-one) or single-shot (one-on-one) experiments. Note that this *incubation effect* was discovered for sources with sub-kHz repetition rates, so the dynamics described by the TTM are fully relaxated when the next shot arrives.

The causes of incubation due to 'precatastrophic' changes in the absorption are manifold. In dielectrics, one can multi-photon-excite single electrons to states in the conduction band. If these do not recombine with the simultaneously generated holes, they form so-called colour centres, which open up 'new' possible excitation pathways for photons from following laser shots. Self-focussing in positive Kerr media can also play a role [92]. The lattice can be affected, in metals as well as in dielectrics, by introducing regional defects, or by micro-boiling of a small volume [93]. For nanosecond pulse damage on metals, Jee *et al.* have classified a range of surface effects induced by local thermal stress-strain [94]. Within the 50% probability range between non-damage and certain damage, they describe many morphologies, starting with heterogeneous removal, or 'cleaning', of the surface, continuing over single slip-line defects and ripple formation, to local flat melting and boiling, at successive average threshold values.

More importantly, in the above mentioned work a phenomenological model is introduced, connecting single pulse damage to the multi-pulse case, where all of these transient states add up noticeably. Encountering the cleanest possible surface, the threshold for single shots $F_{th}(1)$ is highest. For larger shot numbers N, the threshold fluence is modified by

$$F_{th}(N) = F_{th}(1) \cdot N^{\xi-1}. \tag{2.48}$$

As we have stated, if there is any accumulating effect at all, the multi-pulse threshold will be decreased, so that ξ is a number $\lesssim 1$. The case of $\xi = 1$ describes a material where no incubation is present. From Eq. (2.48) it follows that

$$\log(N \cdot F_N) = \log(F_1) + \xi \cdot \log(N). \tag{2.49}$$

In this presentation, the slope of threshold fluences over different shot sizes yields the incubation parameter ξ. Using this evaluation, for example, Jee *et al.* have categorised the vulnerability of different crystal orientations to accumulative shot damage, even for the same material and reproducible initial surface conditions [94].

While this model is helpful for typical shot numbers into the few thousands, there is no valid prediction for continuous exposure, or an infinite number of pulses. It is, however, clear that a fluence value must exist which is *not ever* large enough to induce noticeable damage on the material. Another fit model, proposed by Ashkenasi *et al.*, delivers $F_{th}(\infty)$ as a fit parameter [92]. The evolution for multiple shots follows

$$F_{th}(N) = F_{th}(\infty) + \left[F_{th}(1) - F_{th}(\infty)\right] e^{-k(N-1)}. \tag{2.50}$$

Obviously, the single shot case is retrieved for $N = 1$. Here, the parameter characterising the strength of incubation is given by $k \geq 0$. This model is useful for purposes of this work, as a permanent exposure of material is aspired.

Evaluation of threshold fluences

The technical aspect of material modification using lasers rests on the ablation of molten and evaporated material. Relevant modification is usually defined as permanent visible change in the structure of the material. The evaluation method of ablation thresholds used in this work relies on a *post mortem* analysis of the exposed sites. This means that, after exposure, the diameter D of affected regions is compared to local energies within the laser beam profile. Some, like Jee *et al.* mentioned above, interpolate 50% damage probability values between the onset of modification and certain damage from microscopic evaluation of damage sites. For *in situ* monitoring, it is also possible to view the reflectivity of an exposed site with an integrated setup [95]. This method is found to be very sensitive to the onset of surface modification, however there might be a discrepancy with threshold fluences determined through visibly affected material. We describe the ablation threshold characterisation by area or diameter regression, as it was used as a basis for most of our experiments [94, 96, 97]. It will be shown in the experimental chapter that this easily obtained figure is also useful for (nano-) structured targets.

The energy distribution of a (temporal and spatial) Gaussian focus on a target is calculated from the intensity as [98]

$$E(r) = \int_{-\infty}^{\infty} dt\, I(r, t) = E_0 \exp\left(-\frac{2r^2}{w_0^2}\right), \tag{2.51}$$

where w_0 is the usual $1/e^2$ diameter of the beam. If the pulses are near the ablation fluence of the exposed material, only an area within the focal spot, up to a diameter $D = 2r$ where the threshold is reached, will be affected, and ablation will be visible. The threshold energy is $E(D)$, and from Eq. (2.51) follows that

$$D^2 = 2w_0^2 \ln\left(\frac{E_0}{E_{th}}\right). \tag{2.52}$$

While in most experiments the pulse energy is recorded, the fluence calculates as

$$F_{\text{th}} = \frac{2E_{\text{th}}}{\pi w_0^2}.$$

(2.53)

In a typical evaluation the squared diameters—averaged over a number of identical shot sites—are plotted against the logarithm of the laser energy. In this presentation, the linear fit is offset by the threshold energy, while the focal area is incorporated into the slope. The threshold fluence of the exposed material is obtained from the extrapolation of the fit to $D = 0$ (no visible crater), and by using Eq. (2.53).

Let us summarise all sections of this chapter by their relevance to the aims of the present part of this work, leading to the experimental campaign of investigation. First, we have described the usefulness of extreme ultraviolet light sources, and the advantages of using High-Harmonic Generation for this purpose. The principle of HHG has been explained, historically beginning with a strong field model, and outlining a quantum-mechanical realisation of this model. From an experimental point of view, a typical setup—including macroscopic effects accessible via simple tuning parameters—was described. The necessity of high laser intensities as a prerequisite for HHG, while at the same time aspiring to higher repetition rates, has triggered work on combining strong field optical physics with enhanced fields from plasmonic structures.

We have shown the effect of nanoscopic metallic particles on impinging fields. Resonant excitation of such particles, comparable with large scale antennas, combined with sharply defined geometric features, is commonly used to obtain local high field strengths, which are discussed to be sufficient for HHG in the vicinity of such antennas. The most promising geometry combining these desired properties, the bow-tie, has been broken down into its defining parameters by their effect on resonance and field enhancement. The discovery that employing bow-ties for boosting fields to HHG strengths leads to interesting modifications— first and foremost a cut-off increase in comparison to 'conventional' HHG—was shown, citing current theoretical work on this field.

Still, the natural limit of this concept is set by the damage of metallic objects when exposed to intense laser radiation. The physical mechanisms of absorption and redistribution of energy impinging on a metal surface have been explained. A temporal breakdown of consecutive effects, and some peculiarities that transcend the idealised model conditions, have also been mentioned. Last, we have described the procedure of measuring damage threshold fluences in a manner that is useful for our experimental goals.

Chapter 3

Laser-induced Damage Measurements
on Nanoplasmonic Materials

Aspiring to get a handle on metallic nano-structures that are in theory capable of performing experiments such as the one motivating the current part of this work, we have conducted a number of experiments leading up to the final object of investigation. The very first step was done by examining pure, solid layers of gold. From the beginning, gold was chosen as a plasmonic material since it offers very good properties according to the criteria from Eqs. (2.27)–(2.28), as shown in Table 2.1. Additionally, it is most regularly and favourably used in the fabrication facilities which are at the disposal of the provider of our bow-tie samples. Also, compared to silver as a competitive plasmonic material, gold is not as prone to oxidation, which makes the samples more flexible to handle.

Although solid layers of metals have been extensively researched in their response to ultrashort laser exposure [95, 97, 99], it was important to us to know the properties of the layers that were produced with our specific supplier's tools. This also gave us the opportunity to test our system and laboratory equipment as a setup for determining damage thresholds. With comparable sources in mind, our results from this first measurement could be held up against previously published numbers, while the explained variance due to hardly reproducible conditions could be evaluated for our setup. The thicknesses of tested layers ranged down to approximately the height that we calculated for desired bow-ties. The results and interpretation of the obtained data are presented in section 3.1.

In a next step we treated various nanoscopic objects, emulating the properties of bow-tie gold structures. This means, firstly, that the spatial extent of particular objects was around 100 nm. While in the preceding step layers of thicknesses below that size were evaluated, the one-dimensional model described in section 2.5 does not account for transverse dimensions in this region. Consequently, it is important to obtain actual measured data for LID in the nanoscopic regime. It was of special interest in this step to separate between similar sized objects that were or were not explicitly resonant to our exciting laser wavelength. An according distinction was made and characterised. Additionally, the influence of field-enhancing properties on LID was anticipated by using rectangular, resonant structures with distinct corners. We shall show results for resonant and non-resonant nano-objects in section 3.2, along with a comparison of both cases.

Finally, an exemplary bow-tie sample was examined. We shall break down the initial design, fabrication details and subsequent characterisation in section 3.3. The laser system and attached setup used for all upcoming experiments is described in detail in appendix A.

3.1 Laser-induced Damage on Solid Gold Films

From a first estimate, the height of resonant bow-tie antennas had to be somewhere between 50 and 100 nm for realisable lateral extents. Starting from the simplest assumptions with the depth-dependent model for energy relaxation upon exposure, we first wanted to explore the characteristic length scale for penetration of electrons. For that purpose we examined exemplary gold layers with thicknesses of $\sim$ 30, 80, 170, 250 and 340 nm. The layers were directly sputtered, without the addition of an adhesive layer, onto SiO_2 substrates. As described in section 2.5, threshold fluences were determined by the extrapolation of the diameter of several visible damage sites from specific pulse energies, down to no damage. In order to predict damage under permanent exposure, this was done for 1, 10, 100 and 1000 shots, evaluating the incubation behaviour. For our relevant cases, we only evaluated data for the shortest achievable pulses of $\sim$ 30 fs. We qualitatively cross-checked our results for slightly longer pulses, in case of small maladjustments of the pulse compression in the laser system, or uncompensated dispersion on the way to the sample [100]. However, there was no observable influence on the damage, as this regime is still far below medium and long pulse damage in the greater picosecond region, where pulse-length-dependent behaviour sets in. The results are shown in Fig. 3.1.

The first striking feature of this presentation is the predicted behaviour of the damage threshold for thicknesses greater than 80 nm. For each particular shot number, the threshold fluence remains constant for the respective layers, within errors. The single shot measurements are the least reliable, in terms of uncertainties, since only the average pulse energy in the experiment is known, and not the one actually applied for a particular measurement. In combination with the hardly predictable build-up of the Pockel's cell (see appendix A), the resulting damage sites have a much greater variance than those in the multi-shot case. However, in spite of the resulting large errors, the depth behaviour is particularly well-resolved when single shots are used. This shows that the penetration depth of electrons excited by our applied pulse lengths lies between 40 and 80 nm.

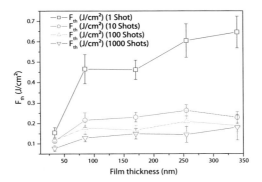

Figure 3.1: Laser-induced damage thresholds of solid gold layers in dependence of layer thickness and for exposure to different numbers of 30 fs pulses.

While thicker films essentially behave like bulk material, the damage threshold for the thinnest, 34 nm layer of $\sim 0.1\,\mathrm{J/cm^2}$ is nearly identical for all shot numbers. When ballistic electron diffusion is inhibited by the bottom end of the layer, the resulting accumulation of energy at the interface between layer and substrate must relax in lateral directions, and back up to the surface. The resulting damage sites are larger in diameter, and the greater volume damage hides evidence of potential surface effects. This means that, opposed to damage in bulk-like material, there is a striking absence of significant incubation behaviour in the thinnest layer. We shall discuss incubation in more detail below.

With single- and multi-pulse ablation thresholds of around $0.6\,\mathrm{J/cm^2}$ and $0.2\,\mathrm{J/cm^2}$, respectively, we correspond reasonably well with comparable measurements from [97] and [99]. Krüger *et al.* fit their obtained thickness dependence to an electron penetration depth of 180 nm, although with a large scattering of data points. An increase of F_{th} above 150 nm can be argued to be visible for our data as well, however the more significant drop happens as determined above, for our setup. We do not consider this to be a significant discrepancy in our respective findings.

The half magnitude of higher damage threshold of bulk-like layers exposed to single shots, as opposed to multiple shots, indicates a strong incubation behaviour. The decrease in F_{th} with ever higher shot counts also holds true for almost all data points. Hence we examined our data according to Eq. (2.49). The results can be seen in Fig. 3.2.

The fit of the linearity $\xi \cdot \log(N)$ proves reliable for the three values for shot counts $10 \ldots 1000$, while the single shot value deteriorates the fit significantly. The large error

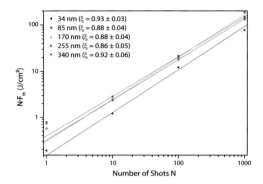

Figure 3.2: Determination of the incubation parameter ξ of Laser-induced Damage on solid gold layers, from the data presented in Fig. 3.1, evaluated using Eq. (2.49).

bars in Fig. 3.1 justify the neglect of these measurement points for the fit, while the extrapolated point of support should represent an acceptable mean value for $F_{th}(1)$. The obtained values of ξ, listed in Fig. 3.2, are seen to be identical within fit errors. Although the majority are greater than those in [97], they concord rather well with findings in [99].

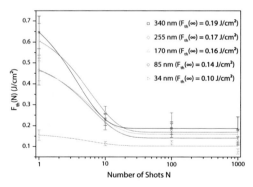

Figure 3.3: Evaluation of the incubation of Laser-induced Damage on solid gold layers with the use of Eq. (2.50), in order to extrapolate for $F_{th}(\infty)$.

In order to find damage thresholds for continuous exposure, we evaluated the same data again, but fitted it to the model given in Eq. (2.50). As was explained, the value of $F_{th}(\infty)$ occurs as a fit parameter in the equation. The results, including the fitted parameters, are shown in Fig. 3.3. All values are determined, by the quality of the fit, to approximately $\Delta F_{\mathrm{th}}\,(\infty) = \pm 0.01\,\mathrm{J/cm^2}$. The results are seen to be hierarchised with layer thickness, showing that, even for a fully incubated layer, there still seems to be an effect of electron penetration in relation to the layer. The levelling to constant behaviour for bulk material probably would only be seen for even thicker layers, see for example [95].

So far, we have determined the pure material damage threshold of thin layers of gold, our material of choice for field-enhancing nano-antennas. Of course, intuitively one would expect the irradiation of nanoscopic particles to be much more destructive. The following experiments were done to quantify this discrepancy.

3.2 Laser-induced Damage on Nano-Structured Gold Targets

The size of structures below 100 nm can be called nanoscopic. We examined particles of volumes similar to those of desired bow-tie antennas. In order to discriminate between pure size effects and energy accumulation due to resonant excitation, as well as field-enhancing effects, we divide this section to examine non-resonant and resonant structures separately.

Damage thresholds of a sample of non-resonant nano-spheres

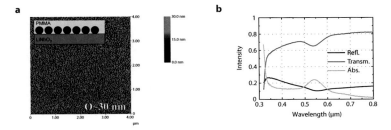

Figure 3.4: Characteristics of the tested nano-sphere sample. **a**, AFM picture of a region of the sample reveals a loosely packed area of arbitrarily scattered, non-stacked spheres on a substrate, e.g. LiNbO$_3$. Elevation levels show that the spheres are between 20 and 30 nm in diameter. After this measurement, the spheres were encased in a protective PMMA layer. **b**, (Simulated) optical characterisation of the global sample shows a broad resonance around 540 nm, but no spectral feature at the laser wavelength of 800 nm (AFM image and simulation courtesy of J. Petschulat).

The tested samples are characterised[1] in Fig. 3.4. At our disposal were three identically

[1]Optical characterisation of resonant samples in all instances of this work, such as in Fig. 3.4 **b**, evaluates prominent spectral features either in the reflected (R) or transmitted (T) fraction of intensity upon broadband illumination. Including dissipative absorption A, these quantities fulfil $R + T + A = 1$.

fabricated samples of the type described in the figure, differing only in that one was produced on a silica (SiO_2) substrate, and two others on lithium niobate ($LiNbO_3$). The protective poly-methyl-methacrylate (PMMA) layer surrounding the deposited spheres is often used to seal and fixate delicate samples because it is fully transparent for most optical wavelengths and only has a slight effect on the overall resonance of embedded metallic objects. The dielectric damage threshold of the PMMA—in the range of $\sim 1\,J/cm^2$—is also much higher than that of the metal, so the observed LID should originate from the covered spheres [101].

We tested the sample for shot numbers 1, 10, 100 and 1000 on respective sites. Similarly to the previously described experiment, we started the evaluation by globally looking at produced holes and using diameter regression to extrapolate for the no-damage case. The results are shown in Fig. 3.5. In comparison to the results for solid gold layers, there is

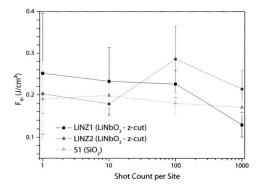

Figure 3.5: Laser-induced damage thresholds of a single layer of $\varnothing \sim 30\,nm$ nano-spheres in dependence of shot number. Three samples with different substrate materials were evaluated.

no discernible difference between single- and multi-shot damage and no influence from the substrate, within errors. Depending on the quality of particular samples, these are quite large, however the 'best' sample (#3 in Fig. 3.5) with the most reproducible damage response can be safely described as lacking incubation behaviour. When averaged over all data points, a damage threshold fluence of the sample as a whole was determined to be $F_{th} = (0.20 \pm 0.04)\,J/cm^2$.

Qualitative comparison with the measurement conducted on solid gold layers reveals that, at a thickness of 30 nm, same as the average diameter of the spheres, threshold values are of the same order or even slightly lower ($F_{th} \approx 0.15\,J/cm^2$, see Fig. 3.1). This

is in accordance with the model that describes the vulnerability of a layer by the relation
of its thickness to the penetration depth of electrons in vertical direction. This quantity
should be the same for the identical pulse lengths used in both our experiments, neglecting
higher order geometric scattering effects inside of the spheres. The fact that we obtain
slightly higher thresholds for damage on the more delicate spheres than on the solid films
can be attributed to the $\sim 50\%$ fill factor of gold spheres making up the sample. Naively,
this means that a smaller portion of energy which would contribute to damage is actually
deposited into metallic material in this case.

The obvious absence of an incubation behaviour can be explained by the probable
damage scenario. We state that, once the damage threshold is reached for *one specific*
gold sphere, destruction of this object by melting and evaporative ablation is complete.
We assume that there is no interconnection between the destruction of single spheres, so
there is no collective surface that could contribute to an incubation effect. To support this
claim, we took a microscopic look at an exemplary damage site, using Scanning Electron
(SEM) and Atomic Force Microscopy (AFM). The results are shown in Fig. 3.6.

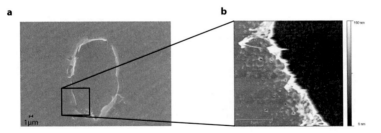

Figure 3.6: Microscopic evaluation of a damage site on the sample of nanoscopic gold spheres.
a, Image of an exemplary site under an SEM. Inside the shot site, the substrate is laid bare, while
on the outside the dielectric PMMA layer remains intact, resulting in the low contrast of the SEM
picture. In the rim region, the PMMA is seen to be lifted off and rolled up. Diameter evaluation,
which was done with a light microscope, is however perfectly possible. **b,** Closer examination of the
transition region by AFM first shows a clear edge between damage and no-damage. However, outside
the crater, there seem to be peripherally decreasing numbers of single circular wells, apparently
reaching the substrate level.

SEM is not really suited for this task for the described samples, since the visible surfaces
are all dielectric. What can still be seen is the frayed edge of the PMMA cover rolling
up along the edge of a damage site in Fig. 3.6 **a**. We interpret this, in accordance with
the described higher damage threshold of PMMA compared to solid gold, as the metal

being affected underneath the layer, which is then ablated along with the evaporating gold particles. In the rim region, this is less pronounced than in the centre, so that some of the covering PMMA remains in place but is detached. In any case, this measurement does not call into question diameter values obtained by light microscopy.

On the other hand, closer inspection of the rim of a particular damage site by AFM shows, apart from the clear-cut edge between the ablated and the remaining material, a number of deep, circular wells with radially decreasing occurrence around the actual crater. We attribute these to single melting spheres that manage to pierce the embedding layer, leaving behind holes that appear to reach down to the substrate. Compared to the original sphere density (see Fig. 3.4 **a**), only a small number of spheres are prone to this damage reaction, whereas all particles are removed from inside the macroscopic crater. This shows that it is problematic to rely on a determination of F_{th} of the sample by evaluating ablated sites by light microscopy only. We conclude that, in combination with a protective effect by the PMMA encasing, it is apparent that there is a significant variance in the reaction to LID by single spheres. As far as the integrity of single metal particles is concerned, our obtained value of $F_{th} = 0.20\,\mathrm{J/cm^2}$ is most likely an overestimate.

Damage thresholds of a sample of resonant, field-enhancing nano-rectangles

a

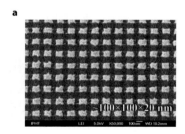

b

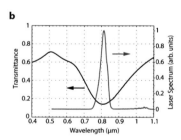

Figure 3.7: Characteristics of the tested sample of resonant, field-enhancing nano-rectangles. **a**, The sample is produced by Electron-beam Lithography out of a 20 nm thick gold layer. The sides of a single rectangle are both approximately 100 nm long, positioned at a 200 nm periodicity (from [102]). **b**, The dip in transmittance of light polarised parallel to the slightly longer axis reveals a resonance that matches perfectly with our employed laser spectrum. The resonance of the shorter axis lies near by, around 710 nm (Optical characterisation courtesy of J. Petschulat).

We shall now compare these results to those from experiments done on explicitly resonant nano-structures. The used sample is shown and characterised in Fig. 3.7. It is an example of a type of structure that is commonly used for Surface-enhanced Raman (SERS) experiments, where the corners of the rectangles are employed to achieve enhanced fields in a confined region [102]. There is no protective PMMA layer used on these samples. Due to the same reasons stated for the nano-spheres, we do not expect an incubation behaviour. In order to make the best use of the available structured area, we only impose single shots of varying energy onto the sample. By the described method of area regression, we obtained a threshold fluence value of $F_{th} = (0.008 \pm 0.002)$ J/cm^2.

Similar to the nano-sphere sample, the effective surface of the resonant nano-rectangle sample—i.e., the metallic part of the perpendicular view of the sample—is only filled to about 25% in this case. This does not necessarily mean that only a quarter of the full pulse energy is applied to the metal, since near-field scattering between the single sub-wavelength structures would have to be considered. However, the trend will be that *less* metallic material is actually exposed to the full fluence inside the laser focus, compared to 'infinitely' extended solid layers. This would also mean that the above mentioned number is an overestimate, as far as the destruction of one single nano-rectangle is concerned.

Let us compare this finding to the materials and structures we have researched before. The depth dependence of damage thresholds has been validated in section 3.1. Extrapolation of the single-shot graph in Fig. 3.1 to layers around 20 nm in thickness, same as the height of the rectangles, delivers a threshold fluence of the solid, pure material slightly below 0.1 J/cm^2, approximately one order of magnitude greater than the value obtained here. Of course, this picture would neglect the nanoscopic nature of the sample. However, F_{th} of the less voluminous—though slightly thicker—nano-spheres is closer to the result for the solid layer than for the rectangles. We conclude from this that the resonance of the nanoscopic object plays a decisive role in its response to the exposure to intense laser light. This might be considered intuitive, since energy is coupled much more effectively into the collective energetic response of electrons in the case of a resonant nano-object. In our experiment we cannot completely distinguish the field-enhancing effect as another reason for greater sensitivity of the samples to LID. However, the enhanced near-field has been identified to drive a relevant mechanism of ablation at sharp features [103–105].

As with the spheres before, we need to get a closer look at the actual damage sites and correlate the microscopic affliction of single nano-particles to the globally evaluated shot sites. SEM images of the nano-rectangle sample after our test run can be seen in Fig. 3.8. In the overview (**a**), we see a rather sharply defined transition in the rim region between the inner, ablated area that was exposed to fluences greater than F_{th}, and the outer area where F_{th} was undermatched. As the blank substrate is perfectly discernible from

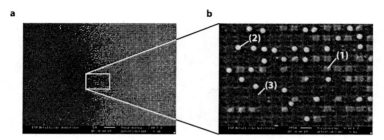

Figure 3.8: SEM images of the transition region surrounding a laser-damaged site on the resonant nano-rectangle sample. **a,** In this magnification, the rim region of the site shows a sharp transition from ablated to intact rectangle (left to right) **b,** Closer inspection reveals three different effects on single nano-objects, namely (1) intact remaining rectangle, (2) rectangle is molten and deformed to sphere, and (3) rectangle is completely ablated.

iridescent reflection from intact rectangles, this transition is identical to the one observed under the light microscope. Thus, the threshold value we determine with area regression corresponds perfectly with the actual damage on the nano-rectangles. In contrast to the previously examined sample, the absence of a PMMA layer makes this measurement very reliable. For samples of this type, the globally obtained number for F_{th} represents a good value to characterise LID, even on the microscopic scale.

We now take a closer look at the transition region in Fig. 3.8 **b**. The following three states can be observed:

(1) The rectangle is left intact at its original site, no degradation is visible.

(2) Material from the former rectangle is not removed, but melted and deformed to a sphere by surface tension. Likely the emerging sphere is detached from the rectangle's former site on the sample. Documentation of this behaviour is also found in [106].

(3) The rectangle is completely ablated.

Seeing that transition from (1) to (2) can occur from one rectangle to the next, the different response is certain to be a property of different rectangles. Fluctuations in laser fluence inside the Gaussian focus would have to be below the wavelength scale of the applied laser. Statistical evaluation, by counting affected sites in a given area, shows that transition from (nearly) full damage to no damage occurs within a stretch of $\sim 4\,\mu$m. Comparing that to the diameters of craters and their respective measurement inaccuracies, this is a smaller contribution to errors of F_{th} than most other uncertainties. This again makes the obtained value of $F_{th} = 0.008 \, \text{J/cm}^2$ rather reliable. It is also interesting to note that, looking again at Fig. 3.8 **a**, it is more likely to find intact rectangles inside the high fluence region of the laser focus than to find afflicted rectangles in a region that was safely exposed to fluences lower than F_{th}. Again, one important conclusion from looking at Fig. 3.8 is that for the non-embedded, free standing nano-rectangle-sample, the evaluated shot site and corresponding obtained value for F_{th} is also valid for the microscopic damage on the sample. Rectangles outside the region in the focus of above-threshold fluence are left intact and, supposedly, functional.

We shall now look at all steps, from production to characterisation of a variety of bow-tie arrays, in the following section. It is to be determined whether the trends in damage behaviour that were explored above can be verified for these intricate structures.

3.3 Measurements with Gold Bow-tie Antennas

For the efficient, reproducible production of metallic nano-structures there are, basically, two competing procedures [56]. The desired patterns can either be written into monolithic material using a Focussed Ion Beam (FIB). This process is advantageous for the quality of the single object. Firstly, the initial material may be produced to a high-grade surface and volume quality using sophisticated sputtering methods. Secondly, ions, with extremely short De Broglie wavelengths, can be focussed to precisely remove sub-nanometre sized chunks of material. However, every single region of the transverse plane of large area structure arrays has to be processed individually. Even with automatised instructions the process is still time-consuming. Kim *et al.* used FIB milling for the production of their samples in the original experiment [29]. The documented pre-enhanced intensities by use of non-amplified oscillator pulses forced them to use very tight focussing, so that arrays of $10 \times 10 \, \mu\text{m}^2$ area, covered with antennas, were sufficient for their purposes.

Electron-Beam Lithography (EBL) represents the competing process. It consists of a series of production steps. The first step is to write a desired pattern into a reactive resist coating using electron beam illumination. This technique is extremely well suited for patterning large areas, as the e^--beam, shaped by a mask, is applied to the whole area at the same time. Regions in the resist that were exposed to the electrons can be removed by chemical means, exposing the substrate. Subsequent evaporation deposition of the metallic material—optionally onto an adhesive layer—fills these holes in the shape of the desired pattern, while the surplus material can then be removed along with the remaining resist by another chemical solvent. Only the free-standing metallic objects remain on the substrate. This can be a slight advantage over FIB-milled structures. As is often the case when using FIB milling, only a small defining volume around the desired object is chiselled away from the solid material layer, as a compromise for greatly reduced processing times. On the other hand, especially because of the chemical steps, but also due to the high demand of precision of the e^--beam and mask, the tolerancing of the singular objects by EBL is somewhat more statistical in nature compared to individual writing by FIB. However, the fast processing of large areas remains its main advantage[2].

When we consider the goal of achieving HHG within the region of a bow-tie antenna, the number of emitters determines the quality of the resulting far-field beam. In this picture, at the position of—ideally—each single bow-tie there is a point-like emitter of High-Harmonic radiation. Because the antennas are excited by the same field, neglecting whether the bow-ties might affect each other, the oscillating enhanced field of each one is in phase. Harmonic light from these fields should thus coherently add up in the far-field. As we were interested, by nature of this work, in the best achievable phase shape of a beam produced by such a scheme, we opted for samples with a large number of exposed bow-ties. This would allow the use of a larger focus area, in accordance with the driving-laser spot sizes, which should thus also relax the restriction on the Guoy phase (cf. Eq. (2.23)) at the site of the interaction. These desired conditions restricted us to the use of our amplified laser, as described in appendix A, rather than aiming for high repetition rates by using a tightly focussed oscillator output. A range of bow-tie geometries meet the resonance of our 800 nm central wavelength pulses while also most realistically allowing the production by EBL. We shall present the design considerations and describe the resulting real antenna arrays in the following section.

Design, fabrication and tolerancing of bow-tie samples

With some initial constraints, such as the realisable height of EBL processed structures, simulations were made to discover possible geometric conditions for the desired resonance frequency of single bow-tie antennas. Suitable for locating the resonance, a fast multipole

---

[2]Recalling section 2.3, we discussed the interplay of antenna length l and gap width g specifically for bow-tie antennas, depending on the production method. It is now easy to see that, when using FIB milling, the antenna is first produced with a fixed l. Later, the gap is cut into the middle, with g determined by the quality of the FIB. On the other hand, when using EBL, two identical triangles are processed at a specific size given by the mask. The triangles are then positioned relative to each other, which then changes l *and* g simultaneously.

method was employed. All simulations were performed using the complex material dielectric function of gold as plasmonic material, and fused silica as a preliminary choice for the substrate with a refractive index of $n = 1.46$. The geometrically constrained parameters were a gap width of 25 nm and a tip rounding radius of 10 nm as realistic values for such a structure. As an arbitrary matter of convenience, the width w (or base) of the single isosceles triangle was chosen to be identical to its height h (or altitude). Both of these mutually dependent parameters, as well as the thickness t of the bow-ties, make up the effective antenna size. Various configurations of $h = w$ and t were tested and the resulting resonance behaviour depicted. The results are shown in Fig. 3.9.

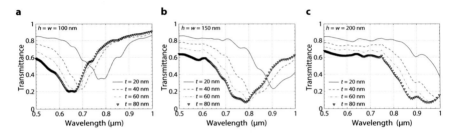

Figure 3.9: Simulated resonance curves of single gold bow-tie antennas on a fused silica substrate. With fixed values for gap size $g = 25$ nm and tip rounding radius $r_{\text{tip}} = 10$ nm, the results from several configurations of $h = w$ and t are shown (**a**, $h = w = 100$ nm, **b**, $h = w = 150$ nm, **c**, $h = w = 200$ nm). With varying t, only the blue curve ($t = 20$ nm) in **a** and, more prominently, the black, magenta and red curves ($t = 80, 60, 40$ nm) in **b** seem to be suitable for resonance at 800 nm (Simulation courtesy of J. Petschulat).

The most promising simulated resonance characteristics were taken as a starting point for EBL processing of large area bow-tie antenna arrays. We chose, by reasons of feasibility, the structure elevation as $t = 60$ nm. The corresponding size of single triangles then had to be $h = w = 150$ nm. However, these parameters were determined under the above stated assumptions, namely on the one hand that the gap size is realised at $g = 25$ nm. On the other hand, the tip rounding for a universal opening angle of the triangles of $\sim 53°$ has no strong influence on the resonance, as was mentioned in section 2.3.

In deviation from the presented simulated data, a sapphire wafer was used as the substrate, since this material is beneficial due to its heat conductivity [107], and thus good dissipation of the deposited energy from the metallic structures [29,38,40]. The validity of the earlier simulations might be compromised since the larger refractive index of sapphire ($n = 1.76$) should have a slight red-shifting effect on plasmonic resonances [108]. We shall address this, in view of resonance measurements on the finished sample, in the upcoming section. An additional layer, 3 nm thick, of titanium, was used to allow for better fixation of the gold to the substrate. It is assumed to not have any strong influence.

The production process is still subject to variation, even for one single employed large area mask for EBL specific to the given geometry. The desired structures are obtained only up to certain tolerances. Two free parameters were allowed to vary in the process. These form the basis for all further distinctions made in this part of the current chapter. Along with their effect on the processing, these are

(1) **Gap size correction**. The single triangle that makes up half of a bow-tie antenna forms the 'unit cell' of the mask. For the full bow-tie, two of these units are used in different orientation. The exact positioning of these two halves determines the resulting width of the gap. Starting at some approximate value, the positions can be changed in steps of some 5 nm. Relative to this initial value, undetermined absolutely, five correction settings were chosen, and are subsequently numbered, in ascending order of the resulting gap size, as $G\{-20, -15, -10, (+)0, \text{ and } (+)5\}$ [nm].

(2) **Electron beam dose**. Even though the geometric path of electrons is generally determined by the mask, the actual dose of the initial exposure of the resist from the e^{-}-beam, before the gold deposition, affects the exact size of the final single structure, i.e., the triangle. For each given gap size we applied five values of electron dose in ascending order. Excluding the mentioned switch in regimes by plasmonic (de-)coupling of the single triangles, they should roughly correspond to increasing resonance wavelengths, and are numbered in the following as $Dos\{01, 02, 03, 04, \text{ and } 05\}$. The actual dose is in units of $\mu C/cm^2$.

The full layout of the sample used in the following is shown in appendix B, in Fig. B.1. As for each given dose the triangle size is fixed, increasing G will result in antennas with a resonance at a slightly larger wavelength, as long as the triangles are coupled by the field in the gap. We have already explained in section 2.3 that, for the largest gaps, decoupling might become an issue, and that the resulting resonance—now determined by the single triangle—will jump to a bluer value. On the other hand, for bow-ties produced with a fixed gap size correction, a low dose (small triangles) might result in two separated—and eventually decoupled—triangles, while a larger dose might lead to touching apices. Even without the fundamental influence of the two free design parameters, the subsequent processing steps can have an additional effect. Especially for larger gap sizes, it may occur that not all of the resist, after gold deposition, is cleared around the metallic structures. This effect can hardly be observed by looking at the resonance, but it may prove to be detrimental for the originally intended experiment, since the gap would be blocked and unavailable to access of target gas. All of these factors result in the fact that, even before testing in an eventual experiment, most configurations are not suited to produce bow-ties of the desired properties. Apart from these parameter regions, shaded red in Fig. B.1, there are settings that have promising outcome in quality of the antennas, shaded green.

Resonance characteristics of the bow-tie antenna arrays

With all of the above described preconditions, we set out to determine the suitability of our bow-tie arrays for high intensity experiments. As was described in the previous section, the position and quality of the resonance seems to influence the damage threshold of the structures. So, without judgement on the existence or quality of the resulting gap and field enhancement, we first classified all parameter configurations for their spectral response. The supposed influence of the parameters on the variations, mentioned above, was to be examined. The results are shown in Fig. 3.10. The reflected fraction of light is used here to visualise the spectral response in a white-light-illumination setup. Reflection from the pure substrate is used for normalisation so that only the isolated effect of the nano-structures is visible. The source was not specifically polarised, such that coupled response of a whole bow-tie should be detected as an additional feature to the single particle response. Each measurement represents a typical area of a given parameter configuration. Even allowing for $\sim 10\%$ absorption in the structures, it is clear that the efficiency of the antenna arrays never exceeds a value of roughly 50%. The rest of the light is transmitted without any noticeable effect from the antennas.

In Fig. 3.10 it is already possible to verify the supposed trend of larger resonance wavelengths for larger e^--beam doses. Also, the overall efficiency of the antennas seems to increase with larger structure sizes. The effects stemming from gap size correction are more subtle. However, especially for the three largest doses, a switch of regimes is visible, where the antenna resonance shows an asymmetry towards larger wavelengths for the smallest aspired gaps. We take this as an indicator that the resonance from large-gap antennas is dominated by the response of the single structure, rather than by a coupled bow-tie. This might be even more problematic, since in these cases the distance to the back-neighbouring triangle could become comparable to the actual intended gap between facing triangles.

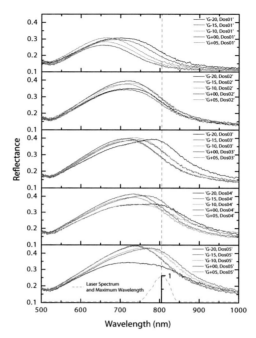

Figure 3.10: Spectral response of the batch of bow-tie arrays, arranged by the processing parameters described above. The spectral response is classified by the reflected fraction of applied white light, normalised for the reflection from the pure substrate. Also shown is the normalised laser spectrum.

We stated earlier that we expect a red-shift of the resonance in comparison to simulations when using a higher-index substrate for the actual sample. Even then, the desired 800 nm resonance is only marginally achieved by the 'best' parameter configurations. We depict the trends for the maximum of the resonance λ_{max}, for all processing conditions, in Fig. 3.11. This illustrates the hierarchical order of λ_{max} with e^--dose. The G-parameter is even more clearly seen to anti-correlate with the expected resonance in most cases.

Damage behaviour in correlation to resonance

We then performed single shot damage measurements on a number of arrays whose resonance curves varied most significantly, judging from Fig. 3.10. As we only had 121 arrays per configuration on the sample, with only enough area to accommodate one single

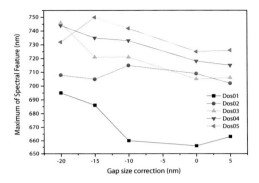

Figure 3.11: Maximum of the spectral response of all bow-tie arrays, grouped by e^--beam dose, smallest to highest. The decreasing resonance wavelengths for larger gap size corrections is a clear trend, implying that in almost all gap size scenarios, the triangles do not form a coupled antenna.

site given by focal spot size, we restricted ourselves to one evaluated shot per selected fluence. We achieve the aspired accuracy by choosing fluence values that were closely spaced instead, since the evaluation relies on the quality of the semi-logarithmic fit over many values of pulse energy (cf. section 2.5). As a dimensionless, qualitative figure of merit classifying the resonance, we evaluated the overlap in the signals of the measured curves from Fig. 3.10 with the normalised laser spectrum (see also Fig. 3.10) as

$$\text{Spectral Overlap} = \frac{1}{100} \int d\lambda\, R\,(\lambda)_{\text{bow-tie}} * I\,(\lambda)_{\text{laser}}\,. \tag{3.1}$$

We show the evaluated data for the threshold fluence F_{th} in Fig. 3.12.

It is immediately apparent that the measured values for F_{th} increase when the main contribution of the resonance of the bow-ties falls outside of the spectral region excited by the laser. As an example, we picked three data points in very different regimes of F_{th} and examined their outcome in more detail. From their resonance curves in Fig. 3.10, rearranged here in Fig. 3.12 **b**, their spectral characteristics especially around 800 nm are seen to differ greatly. Even without the full evaluation of F_{th}, the discrepancy in the respective area's response to LID is clearly seen in Fig. 3.12 **c**. We show three areas of the configurations that were highlighted in **a** and distinguished by their spectral overlap in **b**. These have been exposed to similar pulse energies (210–240 nJ). For comparison, test energies applied in order to obtain a typical single data point in **a** ranged from 130 to 410 nJ. Merely looking at the macroscopic damage visible on the sample, the off-resonant configuration seems to be modified only slightly, in the most intense region of the beam,

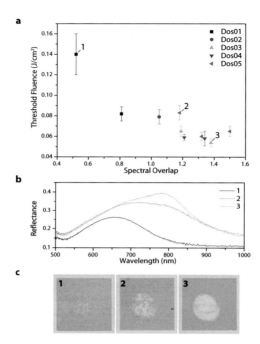

Figure 3.12: Laser-induced damage thresholds of a number of bow-tie arrays, plotted against their overlap in resonance to the used 800 nm laser. **a**, Additionally classified by the dominant design parameter (e^--beam dose, cf. Fig. 3.11), there is a significant trend that antennas with a 'better' resonance are more susceptible to LID. **b**, We illustrate the difference in quality of the resonance by comparing particular resonance curves of three distinct data points marked in **a**. **c**, The quantitative finding from **a** is exemplified by looking at three damage sites, one from each of the numbered types, each exposed to approximately the same pulse energy of ~ 230 nJ.

while the configuration with a resonance peaking near our laser wavelength has a clearly ablated region. We believe this is conclusive evidence that, indeed, the quality of the resonance of a nano-antenna array increases its susceptibility to damage.

The absolute values of damage thresholds of the antennas range from 0.05 to 0.14 J/cm². We recall the characterisation of the resonance of our previously examined spheres and nano-rectangles (see Figs. 3.4 **b** and 3.7 **b**). Our bow-ties with a less than efficient spectral feature are thus located well between the determined values of $F_{\text{th}} = 0.20$ J/cm² and $F_{\text{th}} \lesssim 0.01$ J/cm² for (non-)resonant antennas of different geometries.

Testing the validity of the obtained numbers for F_{th}, we again look at microscopic images of specific damage sites. As in the case of the nano-rectangles, we want to determine whether measuring the diameter of the blurred 'craters' (see, e.g., Fig. 3.12 **c**) evaluated by light microscopy correlates to the actual damage of the nano-structures. We present selected SEM images of typical and peculiar shot sites in Fig. 3.13. Also shown in the respective insets are microscope images as they were evaluated for the data shown in Fig. 3.12. The typical transition region is best seen in Fig. 3.13 **a**. The edge, as visible under the light microscope, marks an actual region outside of which most antennas seem to be fully intact and in place. This substantiates our claim that the area regression method for the determination of F_{th} delivers meaningful results on the single antenna level as well. Inside the shot site, almost all material is either fully ablated, or melted and deformed and, most likely, lifted off the substrate. The latter scenario is cleared up a bit more in **b**. A dark, debris-like region inside the actual damage site is revealed to consist of the scattered remnants of single antennas. While not all parameter configurations were observed to form such 'material dumps', a common scenario for microscopic LID of the bow-ties turned out to include destruction of the antenna configuration, without full ablation and evaporation of the material. We conclude from this that the susceptibility of the contact region (gold–adhesive layer–substrate) to heat deposition is even more critical than the actual material response of the gold particle.

We want to point out some peculiarities in the response to LID by looking at two exemplary shots. Figures 3.13 **c** and **d** document some inner structure of the damage sites that is most likely due to irregularities in the production of the respective arrays. The clear-cut edge found in **c** between destroyed and (predominantly) intact antennas suggests that there was either a problem with the mask or the lithographic illumination step. In contrast, there are more subtle 'wisps' visible inside the crater in **d**, possibly artifacts of the chemical processing steps.

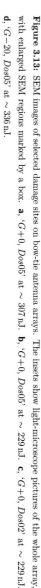
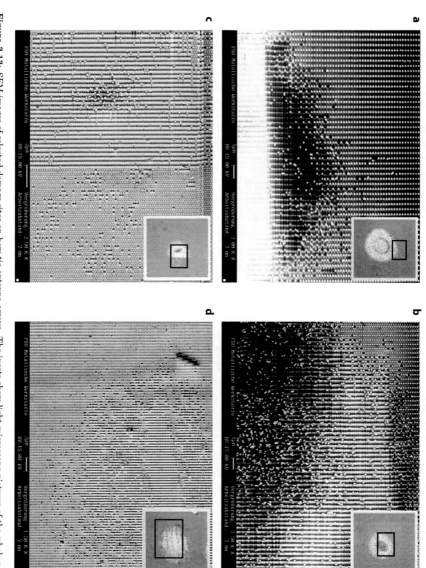

Figure 3.13: SEM images of selected damage sites on bow-tie antenna arrays. The insets show light-microscope pictures of the whole array, with enlarged SEM regions marked by a box. **a**, '$G+0$, $Dos05$' at ~ 307 nJ. **b**, '$G+0$, $Dos05$' at ~ 229 nJ. **c**, '$G+0$, $Dos02$' at ~ 229 nJ. **d**, '$G-20$, $Dos05$' at ~ 336 nJ.

Chapter 4

Summary and Discussion (part I)

This concludes the presentation of our experiments on the damage behaviour of metallic targets, unstructured and processed to nanoscopic dimensions, distinguished by their functionality. We shall now summarise the findings of part I of this work and will conclusively reflect on these findings in the context of other works on the feasibility of HHG employing field-enhanced plasmonics.

First we found the pure material damage threshold of gold, as it would be used as plasmonic material, to be in the lower tenths of J/cm^2, with a significant drop if the perpendicular extent of the material is less than 80 nm. For such thin layers, incubation did not appear to play a role. Thus, for continuous exposure to pulses, regardless of the thickness of the material, a fluence of $0.1\,J/cm^2$ should not be exceeded in order to prevent long term damage effects. With our goal of enhancing fields up to HHG strengths in mind, this value would correspond to a focussed intensity of approximately $3 \times 10^{12}\,W/cm^2$.

Then we determined the damage behaviour of non-resonant and resonant nano-antennas processed from gold, leading up to the actual bow-ties as the original structure of interest. We found that for this task, area regression is still a useful method to determine values of threshold fluence, even though it is originally employed for bulk material only. Strictly speaking, the obtained values of F_{th} thus describe a *macroscopic* sample of the respective structures. This is beneficial for the type of experiment we are aiming for, but is not fully qualified to make statements on the onset of destruction of single nano-structures. Consequently, the observation of the global sample damage was also related to microscopic effects as observed through subsequent inspection of particular structures. The described equalisation of macroscopic with microscopic damage was seen to be problematic when the plasmonic structure is embedded in some way, as was found for PMMA-coated nano-spheres. We addressed the discrepancies which occurred to the inhibition of damage by the coating. The melting of single spheres seemed to be of a more dispersed statistical nature, or it could not be revealed simply by looking at the overall shot site. The evaluated crater is then only the region where the destruction of the spheres also affects the protective layer, as only this is visible under the light microscope. The value of F_{th} obtained from area regression would thus overestimate the threshold fluence for particle damage.

On the other hand, a regular array of free-standing nano-rectangles was determined to exhibit a very clear-cut damage behaviour. SEM pictures, possible due to the lack of a protective coating, showed that the evaluated visible shot site correlates perfectly with the onset of damage on single rectangles. A characterisation of the rim region of affected sites showed that the rectangles melt and evaporate at fluences that vary only slightly from the globally obtained value for F_{th}. This result is encouraging, since eventual bow-tie samples would be produced in much the same way as the present SERS-samples. At this experimental stage it was already implied that the resonance of a metallic nano-structure plays a role in the vulnerability of said structure to Laser-induced Damage. Possible corrections to F_{th} obtained for the spheres, as argued above, would not bridge the gap of one order of magnitude in difference of the threshold fluences between the non-resonant spheres and the resonant rectangles. Corresponding values of the intensities as applied by our laser in the common setup should not exceed $6 \times 10^{12}\,\mathrm{W/cm^2}$ for the non-resonant antennas—about as much as for the thinnest solid layers—and $3 \times 10^{11}\,\mathrm{W/cm^2}$ for the near perfect resonance, in order to prevent damage.

With an experiment as described in [29] in mind, we ventured to examine large arrays of bow-tie antennas, produced by EBL. Even for idealised design parameters, as determined by simulation, the inevitable tolerancing in the production process led to different resonance behaviour, given two free parameters, e^--beam dose and gap size correction. Again, we found that the quality of the resonance plays a decisive role in the response to LID of an array of antennas. As for the sample examined in the present work, only a certain number of parameter configurations led to useful bow-tie structures in the first place, i.e., fully formed, non-touching triangles with a gap width that allowed for plasmonic coupling. Of these, an even smaller number exhibited a resonance behaviour suitable for our laser.

Now, even for the arrays with the best spectral overlap, single shot threshold fluences were as low as $\sim 0.05\,\mathrm{J/cm^2}$. For the typical pulse length of our setup (30 fs, see appendix A), this would correspond to an intensity of $\sim 10^{12}\,\mathrm{W/cm^2}$. We have explained that HHG is an intensity-dependent effect, while the thermal LID of metals depends on fluence. Consequently, by using shorter pulses of a given energy, achieved intensities can be approximately three times higher without affecting the damage behaviour of the structures. While we were not able to determine intensity-enhancement factors of any of our bow-tie arrays, nor use them for field-enhanced nonlinear optical processes, at least a factor of 10 would still be needed merely to reach the onset of HHG without one-shotting the sample. Such enhancement factors can certainly be achieved (cf. Table 2.2), provided that the bandwidth and efficiency of the spectral response of bow-ties were similar, e.g., to the optimised, more primitive structures as represented by the nano-rectangles (see Fig. 3.7 **b**). Then again, as the measurement on those structures revealed, keeping below the respective damage threshold fluence would necessitate intensity-enhancement factors greater than 100 to enable HHG. A much longer design and optimisation process would certainly be needed to fulfil all of those demands on possible future bow-tie samples.

As was mentioned in the introduction to part I of the present work, a lot of scepticism was present in the community as to whether the presented scheme of nano-plasmonic-enhanced HHG were reproducible to the specifications that were originally given by Kim *et al.* While researchers from the same group pointed out that better results were obtained from wave-guide funnels as field-enhancing structures [79], they also reproduced spectra [41], similar but less pronounced than in the original publication. However, they conceded that the employment of bow-tie antennas was subject to long term damage mechanisms that rendered the samples useless after a few minutes. This was even true for illumination of the bow-ties with light of intensities of $\sim 10^{11}\,\mathrm{W/cm^2}$, or a fluence of $0.002\,\mathrm{J/cm^2}$. Similar observations were made for rod-type antennas in [80]. The delayed destruction of the samples at fluences that are significantly below our determined values for single shot catastrophic LID suggests that a mechanism similar to the one described in [103] might be responsible. This is introduced as a high-field-sensitive, cumulative ejection of material, especially at sharp features of field-enhancing structures. A long-running source built according to the original scheme, even if operated under conditions preventing LID as researched in the present work, is apparently not easily realised.

Beyond that, Sivis *et al.* asserted that the only XUV signal they were able to detect, using a very similar setup for their evaluation, more closely resembled undirected enhanced atomic line emission (ALE) than HHG [39]. In a more refined study, they comprehensively compared results from bow-tie-enhanced emission from noble gas atoms with conventional HHG, using estimated numbers and experimentation [40]. They conclude that the sum of the point-like regions of emission from the enhanced-field-regions of the antennas, under the absence of any on-axis phase matching mechanisms, will never be as prominent as the incoherent ALE features they dependably detect. Years after the original scheme has been published, a full evaluation of the potential of nano-plasmonic-assisted High-Harmonic Generation remains outstanding.

Part II

Transfer of Phase Singularities in an High-Harmonic Process

Chapter 5

Theory of Phase Singularities in Electromagnetic Fields

A singularity is in many fields of physics described as a point in space or time where a specific physical quantity is not defined, or not determinable. This usually goes hand in hand with the quantity changing abruptly when passing over the singular point. In most cases, this does not impair a treatment of the problem with the underlying respective model, if special care is taken when addressing the singular nature inside the isolated regions. Nye and Berry have, e.g., looked into dislocations of the phase in general types of wave trains [109]. An immediately obvious and accessible field that was impacted by their findings lies in optical waves and light, where singular behaviour had long been phenomenologically described [110].

Retaining the isolated nature of the singularity, it is perfectly possible to describe optical fields as solutions to Maxwell's equations where abrupt jumps in the phase occur. The singularities in each case are classified by their dimensionality. A line (confinement in one dimension) through a transverse cut of a propagating optical field may be defined to separate half planes where the optical phase, at a given moment in time, differs by a certain value. Similarly, a point-like singularity (confined in two dimensions) with a surrounding spiralling phase evolution is conceivable. The phase at the position of the dislocation is indeterminate in both cases. However, this problem can be covered with conventional electrodynamics, provided that the real and imaginary parts of the field vanish at the singular points, i.e., the field amplitude is zero. The fact that these dislocations turned out to be experimentally accessible makes these field configurations more than simple theoretical oddities. And especially in the case of the point-like singularities the implications proved to be of extreme interest.

In allusion to the spiralling motion of massive particles, the screw-like phase dislocation around an isolated singularity became known as an *Optical Vortex* (OV). We shall give a brief description of its properties in section 5.1. A special case of this phenomenon exhibits the astounding feature of introducing *Orbital Angular Momentum* (OAM) into an optical field and light beam. A theoretical justification of this behaviour and its repercussions, especially on atomic spectroscopy, are given in the following two sections 5.2 and 5.3. By definition, the OV is a feature of the spatial phase of an optical field, and as such was discovered to be of interest in the field of nonlinear optics, and in our special case of HHG. Tying in to the topic of this thesis, HHG is seen to be strongly impacted by control of the spatial optical phase of its driving field, see e.g. section 2.1. As HHG opens the way to short-wavelength, coherent light sources, spectroscopy with light beams carrying OAM would greatly benefit from its versatility. We recount some of the findings of OAM in nonlinear optics in a final section of the theory chapter of part II. The experimental chapter of this part describes our production of an High-Harmonic-generated XUV beam with, for the first time observed, OAM and our evidence thereof.

5.1 Optical Vortices

Let us delve a bit deeper into one of the special cases of optical singularities classified above, namely the point-branch singularity. Given is a specified point in a light field around which the phase of the wave has a spatial variance, continuously depending on the planar polar angle. Graphically, this corresponds to a phase spiral around the point. Mathematically, the singular nature is seen in the depiction of the phase in polar coordinates,

$$\Phi\left(r,\phi\right)=r\cdot\mathrm{e}^{\mathrm{i}\phi}. \tag{5.1}$$

At the origin $r = 0$, the phase ϕ is indeterminate, as the mapping from Cartesian coordinates at the origin ($\phi = \arctan\left(y/x\right)$ for $x, y \to 0$) illustrates. For the electric field to be consistent, this means that real and imaginary parts must be zero at the origin, and thus the intensity exhibits a dark spot. The characteristics of intensity and phase are sketched in Fig. 5.1[1].

The very special case—that makes this field configuration an actual Optical Vortex in the most commonly used sense of the word—occurs when the jump along the branch happens in a multiple integer of 2π. Since the electric field is fully symmetric at a periodicity of the phase of 2π, the 'branch' loses its distinction, because now the phase can be effectively described as continuous everywhere except at the origin. The integer number l denotes the number of times the phase is wrapped around the central axis, or,

---

[1]In fact, the defining criterion for an OV is the phase shape, which only necessitates in a field zero at exactly $x = y = 0$, the model 'point vortex' [111,112]. The importance of the doughnut-like mode depicted in Fig. 5.1, which also fulfils all OV criteria, will be made clear shortly.

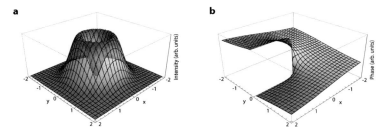

Figure 5.1: a, Intensity and **b**, phase of an Optical Vortex.

mathematically put, for all closed curves C encircling the singularity,

$$l = \frac{1}{2\pi} \oint_C d\Phi. \tag{5.2}$$

It is often called the 'winding number' or Topological Charge (TC). Its implications will be made clear in the following section. From now on, we shall talk about these 'proper' Optical Vortices exclusively, unless otherwise noted.

The step from a rather academic postulation of the phenomenon to an actual realisation by various schemes is very revealing as to the nature of OVs. From the obvious feature in spatial phase, illustrated in Fig. 5.1 **b**, it seems most straightforward to introduce a helical or spiral-staircase-like retarding element into a beam. Implementation has been shown, even for short wavelengths, which poses a high demand on the accuracy of the fabrication [113, 114]. However, this only accounts for the prerequisite of a helical wave front, unless the phase jump at the branch is specifically chosen as $l \cdot 2\pi$ while manufacturing the optic. Even then, the phase shaping property is strictly valid for a single wavelength only.

Holography is another way to access the real-space phase of a beam [112, 115]. Holographic reconstruction is achieved by manipulating the beam in the Fourier plane of an optic. Assume that the manipulation is chosen such as to reproduce the interference pattern of a slightly tilted plane reference wave with the wave scattered off an object. Then one performs the reverse transform by recollimating the beam, modulated by the desired, artificial interference pattern. The phase of the reconstructed wave then looks exactly as if it were projected from the original object. In the case of an Optical Vortex, the 'object' or desired phase shape is given by the above described screw. The holographic interference pattern shows the singularity as a branching point of interference fringes (see Fig. 5.2). For every integer number of Topological Charge l, one fringe splits into

$l + 1$ fringes. Reconstruction from the properly obtained fringe pattern is achieved by

Figure 5.2: Theoretical holographic interference pattern of a beam with a singly charged singularity (from [115]).

inverting the process, using a stepped 2-bit-version of Fig. 5.2 as an interference grating in non-zero diffractive order [112]. The $l \cdot 2\pi$ periodicity of the phase for a given wavelength is automatically fulfilled, or can be adjusted by variance of the incident angle. Furthermore, in a dispersion-free arrangement, this scheme works for spectral distributions, as they would be needed for OVs in short pulses [116].

One seminal realisation, however, came about by figuring out the decomposition of special modes that fulfil the prerequisites of making up an Optical Vortex. Allen *et al.* traced the parallels of such modes in different representations, which not only provided a means to produce OV beams directly from lasers [117], but also revealed their property of carrying Orbital Angular Momentum [118]. More will be said about the consequences of this discovery in the following section, but let us first introduce their findings from the viewpoint of OVs as a phase phenomenon.

As was said, a spatial phase evolution like $\exp(il\phi)$ defines an Optical Vortex, leading immediately to necessarily vanishing field quantities at the origin. One class of transverse, cylindrical symmetric modes that also readily incorporate central field zeroes, and allow for azimuthal phase terms, are Laguerre-Gaussian (LG) modes. They consist of a Gaussian background, modulated by radial polynomials of specified order. It can be shown that they are monochromatic solutions to the paraxial wave equation. One representation, in cylindrical coordinates, is given by [118]

$$
\begin{aligned}
u_{pl}(r, \phi, z) = {} & \frac{C}{(1 + z^2/z_R^2)^{1/2}} \left(\frac{r\sqrt{2}}{w(z)}\right)^l L_p^l \left(\frac{2r^2}{w^2(z)}\right) \\
& \times \exp\left(\frac{-r^2}{w^2(z)}\right) \exp\left(\frac{-\mathrm{i}kr^2 z}{2(z^2 + z_R^2)}\right) \boxed{\times \exp(-\mathrm{i}l\phi)} \\
& \times \exp\left(\mathrm{i}(2p + l + 1) \tan^{-1}\left(\frac{z}{z_R}\right)\right).
\end{aligned} \tag{5.3}
$$

Here, C is a constant, and z_R and w are the usual Gaussian propagation parameters, i.e., Rayleigh range and beam radius. The first row of Eq. (5.3) contains the beam spread at distance z from the beam waist, and the transverse modulated part, represented by the associated Laguerre polynomial

$$L_p^l(x) = \sum_{m=0}^{p} (-1)^m \frac{(l+p)!}{(p-m)!\,(l+m)!\,m!}\, x^m. \tag{5.4}$$

The second row has the Gaussian part of the amplitude and the propagational development of the shape of the wave front. Also, there is the tell-tale azimuthal phase dependence, marked by the box, which secures the rotational symmetry of the radial-coordinate-dependent Laguerre-mode and reveals the association of such a solution with an OV. Finally, the Guoy phase shift, with a contribution from the non-trivial transverse dependence, is given in the last row. In the context of Eq. (5.3), the index l stemming from L_p^l is identified with the winding number from the phenomenological introduction. It can easily be shown that the width of the ring around the central zero scales as $\sqrt{|l|}$, while each integer of p adds a radial field zero to the mode. Here we shall mostly consider the trivial, '$p = 0$'-(ring-)modes.

When symmetry is broken in orthogonal directions, as is often the case in laser cavities, the higher order transverse modes are given by solutions including Hermite polynomials $H_n(x) \times H_m(y)$, with integers n and m. The solutions, similar to the one in Eq. (5.3) but in Cartesian coordinates, are the so-called Hermite-Gaussian (HG) modes. Despite being around since the 19th century, the ramifications of a mathematical connection between both, cylindrical and orthogonal basis polynomial systems was only put into a meaningful context by Allen's group in the early 1990s. Not only can a HG mode, positioned diagonal to the Cartesian axes, be decomposed into a sum of orthogonal, axial HG modes [118],

$$H_n\left(\frac{x-y}{\sqrt{2}}\right) \times H_m\left(\frac{x+y}{\sqrt{2}}\right) = 2^{-\frac{n+m}{2}} \sum_{k=0}^{n+m} (-2)^k\, P_k^{(n-k,\,m-k)}(0)\, H_{n+m-k}(x) \times H_k(y), \tag{5.5}$$

but the same Cartesian decomposition *with an additional simple phase shift* can also be represented by a construction of LG modes,

$$\sum_{k=0}^{n+m} \left(2\boxed{\text{i}}\right)^k P_k^{(n-k,\,m-k)}(0)\, H_{n+m-k}(x) \times H_k(y)$$

$$= 2^{n+m} \times \begin{cases} (-1)^m\, m!\, (x+iy)^{n-m}\, L_m^{n-m}\left(x^2+y^2\right) & \text{for } n \geq m \\ (-1)^n\, n!\, (x-iy)^{m-n}\, L_n^{m-n}\left(x^2+y^2\right) & \text{for } m > n. \end{cases} \tag{5.6}$$

Here, $P_k^{(n-k,\,m-k)}(t)$ is in both cases the same, rather simply constructed constant given by a polynomial of respective order, taken at $t = 0$. A cartoonish, mathematically backed interpretation of this similarity is given in [118], reproduced here in Fig. 5.3.

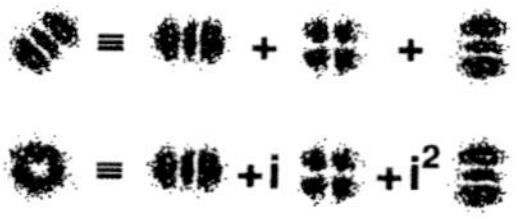

Figure 5.3: Illustrating the mode decomposition of a low-order Hermite-Gaussian mode at 45° into orthogonal HG modes, and of a Laguerre-Gaussian mode into the same, dephased orthogonal HG modes (from [118]).

Making use of this realisation, a mode converter was proposed and implemented, by exploiting the differing Rayleigh ranges for different HG modes [117]. With the proper astigmatic focussing elements, and their respective influence on orthogonally oriented modes, the necessary phase shifts could be effected on the basis modes, and a LG mode could be produced from a He-Ne laser forced into higher order transverse-electric HG operation. However, this clearly visible association of an Optical Vortex with well-defined Laguerre-Gaussian modes allowed another implication, which was previously hinted at. Classical electrodynamic as well as quantum-mechanical treatment revealed the presence of Orbital Angular Momentum in OV beams. We shall outline this fact, along with its applications, in the following section.

5.2 Orbital Angular Momentum in Electromagnetic Fields

Maxwell's theory readily describes energy transport from electromagnetic waves, quantified by the Poynting vector as the directional flow of energy density,

$$\mathbf{S} = \mathbf{E} \times \mathbf{H}, \qquad (5.7)$$

with electric and magnetic fields $\mathbf{E}$ and $\mathbf{H}$, and the momentum density flow,

$$\mathcal{P} = \mathbf{D} \times \mathbf{B}, \qquad (5.8)$$

for the displacement field $\mathbf{D}$ and the magnetic flux-density $\mathbf{B}$ [64]. From a classical mechanic analogy, Poynting himself had concluded that circularly polarised light would exert a torque on a birefringent plate [119]. The plate would have to absorb some quantity, as the beam of light changes its polarisation state, e.g., from left-handed to right-handed circular polarisation, as a matter of conservation of angular momentum. In vacuum, however, Eq. (5.8) cannot yield an on-axis component of angular momentum

$$\mathbf{J} = \varepsilon_0 \int d\mathbf{r} \left[\mathbf{r} \times (\mathbf{E} \times \mathbf{B}) \right],$$ (5.9)

as $\mathbf{r} = (0, 0, z)$ and $\mathbf{E} \times \mathbf{B}$ are parallel for plane waves. Still, Beth was able to measure the predicted effect in 1936 in an astoundingly intricate experiment [120]. Sign and magnitude of the torque on a suspending half-wave-plate corresponded to the change of angular momentum of N photons, each contributing quanta of $\pm\hbar$, when changing the polarisation state. By then, of course, the helicity or 'spin' of the photon and its ensemble average manifestation as polarisation was known. The classical theory was found to be no less inaccurate, considering the full modal description of the light used in Beth's experiment (see footnote [10] of ref. [118]). Conclusively, it was not a large step to postulate higher-order mode constructions that exhibit Orbital Angular Momentum in addition to spin. Putting everything in context, the mentioned seminal work of Allen *et al.* established the role of Laguerre-Gaussian modes in optical OAM.

Staying with the classical picture first, one can calculate the real part of the linear momentum density $\mathbf{M}$ for the paraxial modes introduced in the previous section. In doing this, the momentum flow is now seen to have r, ϕ and z components, as opposed to the strictly propagation-parallel $\mathcal{P}$ of plane waves. Even in the paraxial focus, there remains a finite ϕ-component, spiralling around the propagation axis along z. The angular momentum density can now be constructed using Eq. (5.9). Its time-averaged z-component yields two terms[2] [118],

$$M_z = \frac{l}{\omega} |u|^2 + \frac{\sigma_z r}{2\omega} \frac{\partial |u|^2}{\partial r}.$$ (5.10)

One can see a spin-dependent part in the second term, with $\sigma_z = \{-1, 0, 1\}$ for left-handed-circular, linear or right-handed-circular polarisation, respectively. The former winding number l is now revealed to add l quanta of OAM to the field. This simple additive nature reflects the mutually independent treatment of spin and OAM in quantum mechanics, where total angular momentum $\mathbf{J} = \mathbf{L} + \mathbf{S}$. Here, the ratio of total angular momentum flux to energy flux proves to be [118]

$$\frac{J_z}{cP_z} = \frac{(l + \sigma_z)}{\omega}.$$ (5.11)

---

[2]The structure of the second term reveals that the absence of torque delivered by a pure plane wave is preserved even for circular polarisation, since no field gradient is present. Beth's experiment using a 'real' beam is not contradictory to electromagnetic theory.

Allen *et al.* also showed that this clear separation of orbital and spin contributions is only strictly valid in the paraxial approximation. Barnett from the same group provided the full calculation, starting from Maxwell's equations only [121]. The result shows a mixed dependence of the terms on l and σ_z, while yielding the result from Eq. (5.10) if the approximation is applied only at the end. The role of Hermite and Laguerre polynomials in the solution to the quantum-mechanical harmonic oscillator had already implied that there are similarities between paraxial optics and quantum mechanics. Reference [118] also uses this similarity as a starting point for the investigation of Laguerre-type Optical Vortex modes as carrying OAM. It is a remarkable fact that the paraxial approximation secures l and σ_z as being 'good' quantum numbers of Laguerre-Gaussian modes concerning (spin-) angular momentum operators [122].

The consequences are far-reaching and intriguingly useful from a practical point of view. Experimental realisation and application shed light on the interplay of the quantum-mechanical origin as well as classical, even mechanical manifestations of angular momentum. One can, for example, use an OV beam in a special case of an optical tweezer. In addition to the field-driven positioning of particles in the conventional tweezer concept and the spin transferred from the polarisation state, beams with optical OAM can exert a torque on the object, moving it at an arm length at distance R around the central beam axis [123]. The dark central core that persists in the near- and far-field alike bears an advantage for special trapping geometries, free of radiation pressure and unwanted direct light-matter-interactions. The same concept could even be extended to optically driven microscopic machines [124]. From the quantum-mechanical introduction of optical OAM arise even more fundamental phenomena. First and foremost is the confirmation that OAM is a quantum property of the single photon. In parametric down-conversion, or the 'breaking up' of photons, not only is conservation of angular momentum observed, but also is the resulting two-photon state a coherent superposition of entangled OAM quantum states [125]. This realisation opens pathways to new encoding methods for quantum information processing [126].

As was outlined in the introduction to this chapter, we set out to generate Optical Vortices in photons produced in the XUV, via HHG. Spectroscopy in the XUV [1] would also greatly benefit from the addition of OAM of light as an additional degree of freedom. We want to discuss the origin and implications of this point in the following section.

5.3 Optical Orbital Angular Momentum in Spectroscopy

It has been verified that consistency is retained for the treatment of light carrying OAM, whether by classical electromagnetic theory or at the single photon level. The description of spectroscopy on atoms is, in principle, also consistent in both pictures, with the most commonly adapted model being an hybrid of a quantum-mechanical description of the atomic system, driven by a classical representation of an electromagnetic field. Symmetry considerations emerge as selection rules for the excitation of transitions. These rules gain justification in a full quantum theoretical model, which will only be intuitively formulated here. For example, in a single-photon-stimulated transition, total angular momentum must be conserved. The photon, carrying a single quantum $\hbar$ of angular momentum as its spin (helicity) state of $|\pm\sigma\rangle$, is annihilated in the process. Consequently, the atomic electron is elevated to a higher energetic level. Apart from shell transitions Δn, the angular momentum quantum numbers must change by the same number of quanta $\hbar$,

$$|\psi_i\rangle = |l, m\rangle \rightarrow |\psi_f\rangle = |l', m'\rangle \,, \tag{5.12}$$

with (dipole) selection rules

$$\Delta l = \pm 1, \ \Delta m = 0, \pm 1. \tag{5.13}$$

Now, in accordance with the theory of optical OAM introduced in the last section, it is to be expected that photons carrying additional quanta of angular momentum can overcome classical selection rules [127]. In the cited first treatment of extended selection rules, this is viewed from the perspective of multi-modal fields and the ensuing non-trivial spatial dependence of the vector potential. Higher order fields, for instance including quadrupolar terms, are classically capable of overcoming the 'single-photon' rules from Eq. (5.13). Van Enk described this behaviour as field properties acting separately on internal variables—i.e., electronic 'motion' in a fixed nucleocentric frame of reference— and external variables—i.e., centre-of-mass motion of the atom, much like the optical spanner mentioned before. This leads one to identify these quantities with the postulated independent OAM and spin components of an electromagnetic field.

More recently, Picòn *et al.* have performed full simulations of the interaction of short optical pulses carrying quanta of OAM with model hydrogen atoms [128]. They verified extended selection rules in the transfer of ground state electrons into continuum states. From these results, they argue that the same principle should also hold true for transitions between bound states. For simplicity and proof of principle, they use few-cycle pulses at an optical frequency of $\omega = 1$ a.u. ($\lambda \approx 46$ nm). This wavelength lies perfectly inside the regime of optimal HHG as it is routinely achieved, e.g., with our available laser system on argon. We shall describe some details of this simulation, as it might be considered a trial experiment for XUV light with OAM.

In contrast to the previously cited source, the atom is modelled as rigid, so that only the electronic terms enter the field-driven Schrödinger equation. The full Hamiltonian is constructed with canonical momentum and Coulomb potential as

$$H = \frac{1}{2m} \left(\mathbf{p} - q\mathbf{A}\left(\mathbf{r}, t\right) \right)^2 + qV\left(r\right). \tag{5.14}$$

It is most convenient to separate the Hamiltonian into well-known parts. First, the pure kinetic and Coulomb terms describe the free atomic system with textbook solutions. From the squared bracket in Eq. (5.14) remain the interaction terms,

$$H_{\mathrm{I}} = -\frac{q}{2m}\left(\mathbf{p} \cdot \mathbf{A} + \mathbf{A} \cdot \mathbf{p}\right), \tag{5.15}$$

and the terms quadratic in the field,

$$H_{\mathrm{II}} = \frac{q^2 \mathbf{A}^2}{2m}. \tag{5.16}$$

The latter part of the Hamiltonian, associated with two-photon processes and according selection rules, is of lesser significance than the former. The main part of the line of argument retraced here will concentrate on the single photon selection rules that arise from symmetries in H_{I}.

Now, the initial quantum state is assumed to be the pure ground state of the free Hamiltonian H_0. The electromagnetic vector potential is constructed as a few-cycle pulse with a $\sin^2$ envelope and a Laguerre-Gaussian mode as defined in Eq. (5.3). Polarisation is also included by defining a spatial dependence of the amplitude as $\mathbf{A}_0 \equiv A_0 \left(\alpha \, \mathbf{e}_x + \beta \, \mathbf{e}_y\right)$, with appropriate choice of α and β. The special cases of left- and right-handed circular polarisation are reproduced by choice of $\alpha = 1/\sqrt{2}$ and $\beta = is/\sqrt{2}$, with $s = +1$ (left) or $s = -1$ (right). The interaction happens in the paraxial focus of the beam. The selection rules due to H_{I} emerge from the calculation of the transition probability amplitude $\langle \psi_f \,|\, H_{\mathrm{I}} \,|\, \psi_i \rangle$. Rewriting the linear momentum operator by its canonical transform, and using the commutator of the position operator and the free Hamiltonian, the amplitudes are given by [128]

$$\langle \psi_f \,|\, H_{\mathrm{I}} \,|\, \psi_i \rangle = -\frac{e}{i\hbar} \Delta E \, \langle \psi_f \,|\, \mathbf{r} \cdot \mathbf{A} \,|\, \psi_i \rangle \tag{5.17}$$

with the energy difference ΔE between initial and final state. Both of these states, as free solutions of H_0, can be expressed using spherical harmonics Y_l^m for the angular dependence.

Inserting the LG construction of **A** as described above, the crucial information comes from the integral defined by Eq. (5.17). As an example, the term resulting from the $e^{-i\omega t}$-part, or absorption of one photon, leads to an integral for the dependence as [128]

$$I(l,s) = \int Y_{l_f}^{*\,m_f}(\theta,\phi) \left[\sin^{|l|+1}(\theta)\, e^{i(l+s)\phi} \right] Y_{l_i}^{m_i}(\theta,\phi)\, d\Omega. \tag{5.18}$$

The bracketed part of the operator $\mathbf{r} \cdot \mathbf{A}$ can also be expressed in terms of spherical harmonics [129], and the well-known orthogonality of this basic set of functions applies. Non-vanishing solutions of Eq. (5.18) exist only for [128]

$$|\Delta l| \equiv |l_f - l_i| \leq |l| + 1 \leq l_f + l_i, \qquad \Delta m = l + s, \tag{5.19}$$

where $\Delta l + l$ must be odd. Note here that the emission ($e^{+i\omega t}$) part of the solution to Eq. (5.17) leads to a very similar result, with an inverted change of quantum numbers. This shows that absorption and stimulated emission are treated equivalently.

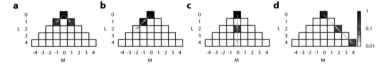

Figure 5.4: Illustrating the selection rules for ionisation of hydrogen with light carrying a certain number of quanta of OAM. Shown is the projection of the final state, after excitation, into spherical harmonics, decoded by their respective orders L and M. The arrows correspond to allowed transitions. **a**, $l = 0$, linearly polarised Gaussian. As linear polarisation is a superposition of both left- and right-handed circular polarisation, Y_1^1 and Y_1^{-1} are dominantly excited. **b**, $l = 0$, right circular ($s = -1$) polarised Gaussian. The right half of the triangle is 'forbidden'. A small component due to two-photon-absorption is visible. **c**, $l = 1$, right circular polarised LG. The value of l can change by 2, or twice that for two-photon processes, but, as $\Delta m = l + s = 0$, only $m = 0$ states contribute. **d**, $l = 1$, left circular ($s = +1$) polarised LG. Only fully additive states ($\Delta m = 2$) contribute (from [128]).

Taken from [128], Fig. 5.4 shows the spherical harmonic components in the final state of an electron removed from an hydrogen atom, after ionisation by photons from a beam as described above. Select cases of l and s of the beam are given in Fig. 5.4, **a–d**.

Let us briefly summarise the connections between the introduced quantities in the respective classical and quantum-mechanical pictures. Higher order terms in the evolution of valid solutions for the electromagnetic vector potential can mediate interactions beyond the dipole limit. Classically, this comes from the non-trivial spatial dependence of the modes. Applied to light-atom interactions, this is expressed in a separate excitation of internal (electronic) and external (orbital motion) degrees of freedom. A special solution of such modes are those of Laguerre-Gaussian type, that also fulfil the $\exp(il\phi)$ criterion that describes an Optical Vortex. In this case, l is a good quantum number for a quasi-classically constructed angular momentum operator. Coincidence measurements prove that all of these properties derived from electromagnetic theory also manifest on a single photon level, such that the photons can be described as carrying Orbital Angular Momentum.

It has been shown that it is possible to imprint OVs on light that is produced in the XUV and x-ray spectral region [114, 130, 131]. These experiments, however, were done on unwieldy undulator sources. To make use of the availability of lab-size HHG XUV sources, we set out to find evidence of OV signatures in an High-Harmonic beam that was generated with helical light. While the spiral phase plate scheme from refs. [114] and [130] could be applied to conventionally generated HH light, the aim of this research was the additional fundamental interest of the stability of an OV under such a highly nonlinear process. Before presenting our own experiments, we shall first conclude the theoretical section of part II of the present work by giving a brief overview of nonlinear optical frequency-conversion phenomena that have been examined in the light of OVs.

5.4 Optical Vortices in Nonlinear Processes

As the simplest nonlinear optical effect, Allen's group attempted Second Harmonic Generation using beams of well-defined LG modes. It is straightforward to expect that the second harmonic $\mathbf{E}^2$ of an LG electric field, at the position of the beam waist, should be constructed as in Eq. (5.3), under transformation of the variables[3], in particular $l \rightarrow 2l$. Thus, conservation of angular momentum is implied, as was explicitly verified soon after [132]. (Parametric) three-wave processes allow addition/subtraction or reversal of Topological Charges, as expected from the formal treatment using complex-valued coupled waves with vorticity [125, 133]. Most other nonlinear effects, simulated or experimental, researched with OV light are based on Four-Wave Mixing processes, for example cascaded up-conversion of OAM via Raman sidebands [134–136]. Intensity-dependent nonlinearities, however, such as Kerr-type spectral broadening and supercontinuum generation, tend to suffer from the peculiar spatial profile of OVs, and susceptibility to fluctuations therein [137].

Frequency conversion processes aside, mere propagation in nonlinear media has been shown to challenge the stability of OVs, especially of high TC [138,139]. These modulational instabilities play a role in the interpretation of our findings in chapter 7.

---

[3]This simple, intuitive treatment is strictly valid only for LG modes with $p = 0$. For cases $p > 0$, the resulting field is a mixing of several LG modes.

Chapter 6

Measurement of an Optical Vortex in an High-Harmonic beam

We are making use of the spatial shaping techniques that have long been put to advantage in our laboratory exploring their influence on High-Harmonic Generation. Summarising the most recent experiments, see for example [140]. The realisation of imprinting the spatial phase signature of an Optical Vortex onto our given fundamental laser beam profile happens more or less in the same setup described therein. We shall give a short overview of the necessary instruments and their implementation in the first section of this chapter.

The beamline for HHG, with the prerequisites as outlined in section 2.1, converted for the present experimental use, will be described in the two immediately following sections, with technical details and figures in appendix C. For reference, a measured typical HHG spectrum with argon as a target gas is shown in Fig. 2.3 **b**. Optical components in the far UV have to account for the peculiarities of light in this spectral range. As transmission of all dielectric materials is extremely low, and as diffractive elements have to comply with the extremely short wavelengths of XUV, the transfer of common detection schemes for OVs is challenging. We present a combination of methods that pin-points the vortex nature of our generated XUV light by looking at tell-tale signatures in the intensity profile in the far-field of the freely propagating beam. The characteristics of the phase front of an OV are investigated by applying one of the simplest interferometric methods of wave front splitting by a thin wire. Also, certain expected features could be ascertained by employing a home-built variety of a Hartmann sensor. Finally, we compare the characteristics of spatial modes in different orders of the harmonic comb. All results and their implications, some of which have been published in [141], are presented in section 6.3.

To our knowledge, this is the first experimental evidence of an OV that has survived a nonlinear process of such high order, being only imprinted on the fundamental beam. In the context of theoretical expectations and the extent of our research, we shall interpret our results in the final chapter of part II of this work.

6.1 Phase Manipulation of the Fundamental Laser Light

The fundamentals of OAM in light beams have been presented here only in the case of monochromatic light. The introduction in Eq. (5.3) is explicitly formulated for the spatial dependence of the phase only. Furthermore, the numerical ultrashort example driving the treatment in section 5.3 is written out in the exact same way, with an artificially constructed time window, neglecting a conclusive spectral representation [128, 129]. As has been mentioned before, dispersion-free methods of producing OVs have been proposed and realised [116]. However, we found that our employed method of strictly spatial phase forming using a Spatial Light Modulator (SLM) delivered satisfying results, if the system is carefully tuned. The necessary iterations to achieve this are explained below, while the effects of correct tuning shall be presented in section 6.3.

The input with which an impinging beam is modulated to carry an OV looks like a grey-value spiral with an arbitrarily oriented phase jump from maximum to zero (see appendix C). While this corresponds to a singly charged OV (TC = 1), higher topological charges n can be realised by 'steepening' the ramp so that additional phase jumps are accommodated at $360°/n$ orientation. Examples are shown in Fig. 6.1, along with the resulting beam profiles.

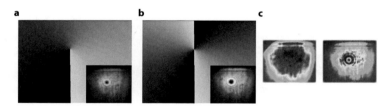

Figure 6.1: Grey-scale patterns applied to the SLM resulting in TC = 1 (**a**) and TC = 2 (**b**) vortices, and the corresponding modes in the fundamental. **c**, False-colour intensity profile of the unshaped (left) and 'TC = 1'-modulated (right) fundamental.

With the SLM model specifically designed for the 800 nm region, there are still two parameters that need to be optimised in order to obtain valid OV modes. These are the positioning of the point singularity with respect to the input beam profile and the exact grey value for a 2π phase jump at our precise carrier wavelength. The first is essential in order to produce a well-defined, stable mode like the one given by Eq. (5.3). In our OV generation scenario, the phase is imprinted separately on an already existing Gaussian mode. An additional intensity shaping effect of the SLM is negligible, and, at worst, identical over the entire beam profile. Only the fully consistent relation of spatial intensity and phase profile, such that the singular point coincides accurately with the centre of the input Gaussian, makes up an OV mode with the expected behaviour. Disregarding this, it is still possible to see a dark spot where the phase is undefined, as in the insets of Fig. 6.1. In propagation, and especially in subsequent focussing, however, the localisation of the singularity on the Gaussian background will be unstable. The vortex core is gradually driven out, along the transverse Gaussian field gradient, ending up in the peripheral wings where the field is close to zero [111]. The effects of faulty placement of the point singularity will be addressed in section 6.3, as they were observed in the actual experiment. For near full use of the active area, the positioning is accurate to $\sim 2\%$ of the full beam width.

The other parameter determines whether the spatial phase is joined correctly at the branch. If the 8-bit grey value is faultily chosen, the branch can be seen in the shaped beam profile as a dim line, as in an imperfect interference pattern. This is readily observed simply by looking at the profile downstream from the SLM, and could easily be experimentally tuned, down to the contrast resolution of the device. With the specifications given in appendix C, the accuracy of the phase modulation step amounts to $\sim 25\,\mathrm{mrad}$, or $\sim \lambda/256 \approx 3.1\,\mathrm{nm}$. For a well-chosen step, the mode looks like the ones in the insets of Fig. 6.1, with no discernible dim line, regardless of the angular orientation of the branch on the SLM.

The insets in Fig. 6.1 **a** and **b** show the resulting modes generated with the respective phase patterns, and observing the two above explained rules. These are photographs of the modes on a piece of paper after about one metre of free propagation across the laboratory table. Both taken at nearly the same spot, one can see the expected widening of the dark core of the OV when the topological charge is increased. Input fundamental modes of the shown orders 1 and 2 are the ones that are used for the actual evaluations that will be presented in later sections. More striking than in the photographs, we also show the false-colour coded intensity profile, before and after phase shaping, in Fig. 6.1 **c**. Apparently, the input mode is only as close to a regular Gaussian as the output of the laser allows, and the pure phase shaping actually results in a point vortex (cf. footnote 1 of chapter 5). The point diffracts immediately behind the SLM, resulting in the concentric Bessel-like rings visible in the modes, as well as a widening of the dark cores in propagation [112]. Somewhat laxly, we nevertheless refer to the resulting modes as '(Laguerre-)Gaussian'. The beam leaves the SLM slightly divergent to optimise focussing conditions in front of the HHG chamber.

6.2 High-Harmonic Generation with Vortex Beams

The beamline and focussing geometry is very similar to the one described, for example, in refs. [19, 20]. The slight deviations, as well as the detection schemes relevant here, are detailed in appendix C. Before every HHG experiment that was performed using the basic setup described here, a newly sealed nickel tube is perforated at its destined site in the chamber, by using the actual driving laser. This is to facilitate overlap by defining the interaction region with the laser itself, instead of finding a pre-drilled opening with the focus. Furthermore, for slight variations in beam quality, focus size and distance or pointing, it is ensured that the opening is exactly as large in diameter as it has to be, in order to minimise gas load in the chamber during operation. However, while for a typical experiment the best Gaussian beam is sought, which always produces the smallest spot sizes, spatial shaping for non-trivial modes such as LG will foremost result in a significant increase of the $1/e^2$ diameter. In the experiments described here, this necessitated the enlargement of the initial hole by slight lateral movements of the focussing optic while exposing the nickel tube. Consequently, slightly lower high-end backing pressures were achievable without imposing too large a gas load on the evacuating turbo-molecular pump.

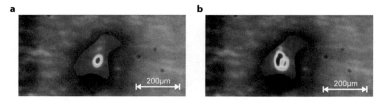

Figure 6.2: Mode profile of the Gaussian (**a**) and Vortex (**b**) fundamental ($\lambda \approx 800\,\text{nm}$) foci inside the interaction region. Overlaid is an image of the laser-drilled hole in the nickel tube, taken outside of the chamber. The respective colour-coded intensities are not to the same scale.

In Fig. 6.2 we show the mode profile at the focal spot in the actual chamber, for comparable HHG by Gaussian (**a**) and a singly charged Vortex fundamental (**b**). The false-colour images were taken using the focus diagnostic, as illustrated in Fig. C.1 **c**. Bypassing the subsequent beamline, a replaceable mirror sends out the fundamental light coupled through the nickel tube. The external optical system is adjusted to image the focus onto a CCD camera. With the absolute image size justified, we overexposed an image of the exemplary shape of the laser-drilled hole, taken outside of the setup with a light microscope. In Fig. 6.2 **b** we see that the intensity zero in the central region is pertained even through focussing. This is to be expected from the inherent phase property of an OV, as opposed to a mere intensity modulated profile that would fill by diffraction [112]. However, the focus does not reproduce an ideal Laguerre-Gaussian case. The expected full circle is interrupted at almost opposite angular positions, and the overall shape is slightly elliptical. We attribute the deviations to the aforementioned impurity of the laser output, illustrated in Fig. 6.1 **c**. However, there are no visible Bessel-like rings around the central mode, suggesting that there is no sharp transverse cut-off in the fundamental mode [112]. The two-fold intensity maxima might, however, result in a filamentary breakdown, once the interacting gas is present [142]. We shall keep these findings in mind when we describe the obtained XUV profile in the following section.

6.3 Intensity and Phase Signature of a Vortex in the XUV

6.3.1 Intensity profile of the beam

This part of the experiment was performed using the beamline described in Fig. C.1 **b**. In the first step, the SLM was fed with a uniform black image, thus acting as a simple reflective optic in the beam path. For an immediate comparison between High-Harmonic Generation in argon with a Gaussian driving beam and one carrying an Optical Vortex, we first iterated the common procedure of optimising HHG yield [18]. This includes scanning backing pressures in the tube and varying the propagational focus position in relation to the centre of the interaction region for best signal (see Fig. 2.3 **b**). Using the found parameters we continuously reduced the laser intensity until the High-Harmonic signal all but vanished. Assuming a well-defined high-energy cut-off for given intensity, and accounting for the low-energy onset of transmission of the equipped aluminium filter $(0.3\,\mu m)$, we ascertained that only one harmonic can be detected. In this way we made sure to observe only the 11th harmonic of $\lambda \approx 800\,nm$ at $\sim 74\,nm$. The resulting XUV profile is shown in Fig. 6.3 **a**.

While leaving the setup completely unchanged, we then switched the SLM instruction pattern to a singly charged OV as shown in Fig. 6.1 **a**. The generation of a valid OV mode was secured by following the steps described in section 6.1. The phase step adjustment is possible by looking at the fundamental mode, while the exact positioning of the Vortex centre is achieved only by evaluating the HHG output, as will be described shortly. As was stated and documented in Fig. 6.2, the spot size of a non-trivially shaped beam is increased over the Gaussian case. Consequently, the laser power had to be deregulated

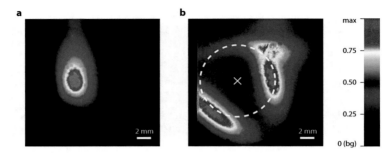

Figure 6.3: Spatial intensity profiles of an otherwise identical HHG experiment, one with a flat phase instruction applied to the SLM (**a**), and the other with an OV of TC = 1 (**b**). With immediate switching back from the latter case to the former, the Gaussian spot appeared exactly at the site indicated by the white cross in **b**. The dashed circle is to guide the eye. Absolute signal varies strongly, so both images have respective background correction and normalisation, with intrinsic scaling as indicated by the colour bar.

from the previous setting, to the point where the HHG signal just begins to appear. We safely assume that the detected XUV light then again comes from a single harmonic. A typically achieved spatial profile is shown in Fig. 6.3 **b**.

The procedure of tuning the fundamental intensity and, accordingly, the HHG cut-off with an absolute calibration given by the onset of aluminium filter transmission should ensure that the High-Harmonic process occurs under identical energetic conditions in both cases, Fig. 6.3 **a** and **b**. However, the signal is drastically decreased in the Vortex case, such that 2×2 pixel binning was necessary to obtain a comparable dynamic. The figures are respectively normalised for viewing convenience. Still, with subtracted background, a dynamic zero is visible in the OV case at the exact site where the Gaussian spot is located with flat SLM. We take this as a first indication that we have transferred the properties of an OV, over a non-perturbative nonlinear process of 11th order, onto an XUV beam.

A full Vortex circle is not clearly discernible, but two opposing 'lobes' of signal, almost equidistant from the core. With the sensitive dependence of the extremely high-order HHG process, a phase matching effect might be the cause of the incompleteness of the Vortex ring. Furthermore, bearing in mind the typically achievable fundamental focus (see Fig. 6.2 **b**), the inferior initial profile of the laser output or imperfect transfer of the focussing system could entail conditions for successful HHG being reduced only to certain regions inside the focus. A break down of an OV beam into filamentary fragments inside a gaseous medium has been predicted and reported [142, 143]. Clearly, further investigation needs to be performed in order to substantiate the evidence of a Vortex surviving HHG.

But first, as an instructive test, we present images of the spatial mode of High-Harmonic signal generated using an OV (TC = 1) fundamental, (de-)tuning the exact placement of the singularity within the Gaussian input. We show these in a suggestive arrangement in Fig. 6.4. The starting point is the configuration, very similar to the one resulting in Fig. 6.3 **b**, where the opposing lobes in the profile produce near equal signals. From this, we shifted the centre of the spiral image over the SLM area in steps of 30 pixels, or 0.6 mm, and show the resulting HHG image. For each one, we use the same intensity scaling, also seen in the figure, accepting clipping at high signals for the sake of visible dynamic.

Bearing in mind that the unmodified laser mode nearly fills the entire active area of the SLM, the singular point never even leaves the most central region of the Gaussian profile. The resulting HHG profiles, however, differ greatly. For one, almost the lowest overall signal results from the ideal—i.e., most equally distributed—'Vortex' case. Now, especially in the diagonal upper-left–lower-right (this corresponds to an horizontal shift on the SLM), we see that the main contribution of the signal switches from one lobe to the other. It should be noted that once the branch point moves even further away from the optimal position, the result is identical to a Gaussian input, as expected. In that case, the overall signal is off the scale of Fig. 6.4. We take this as a clear indication of the Vortex nature of the generated XUV beam, as the characteristic profile depends sensitively upon the validity of the input as a faithfully constructed LG-like mode. If this is not secured, the fundamental Vortex is unstable under focussing, and the Gaussian case is reproduced.

We also verified that the appearance of two distinct off-axis regions of High-Harmonic signal under the described conditions is not just a matter of better phase matching in these regions. We accomplished this by scanning the gas pressure as another method of probing phase matching conditions. As it was laid out in chapter 2.1 of this work, target gas density is an essential factor in matching the phase for HHG in propagation direction. Typically, i.e., when performed with a conventional Gaussian fundamental profile, this factor has to be tuned in relation to the spatial phase in the interaction region. For a vortex mode, the propagational spatial phase is modified, as is seen by the Guoy phase term (third row of Eq. (5.3)) of an LG beam, in contrast to a pure Gaussian. In this case, the intrusion into the phase matching balance by changing the gas density is even stronger, as the slope of the Guoy phase is steeper. In our experiment we observed no change in the spatial distance of the lobes in one particular harmonic order when performing a large

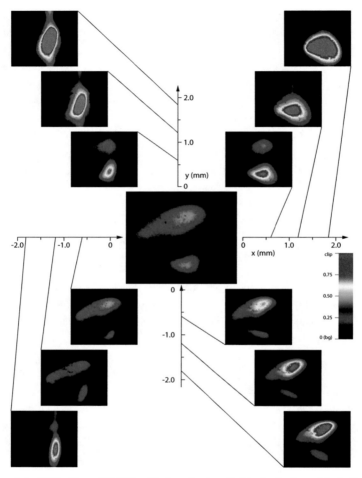

Figure 6.4: HHG with an OV (TC = 1) phase front applied to the fundamental. In the large centre image, the placement of the phase singularity is optimised to produce opposing lobes of equal intensity, within pixel resolution on the SLM. In this case, the overall signal is lowest. Shifting the singularity along two orthogonal axes in steps of 30 SLM pixels, or 0.6 mm, the resulting signal is shown. For the sake of visible dynamic, each image is presented at the same intensity scaling (see bar), in spite of high scale clipping at the outermost configurations.

range scan of gas pressure (see supplementary material to [141]). We *did* observe the typical overall signal variations when global phase matching was adjusted. We take this as a clear indication that phase matching affects the generated (Vortex-) beam *as a whole*, and not just individual segments *inside* the beam, away from the central axis.

Of course, for a full characteristic of an OV, observation of the phase is essential. Interferometric setups for this analysis in the XUV are in principle possible, but suffer from the peculiar optical properties of most materials in this spectral range, as well as intrinsic ambiguities in recovering the phase of an optical field [130]. Thus, it is very demanding to transfer established Vortex detection techniques into the 20–30 nm regime. We describe two possible schemes and the respective results from both in the following.

6.3.2 Detection of the phase signature

Phase evaluation by division of the wave front

One method of phase retrieval that is most easily realised is a kind of interferometry in the Fresnel regime produced by splitting of the wave front [144]. In the originally described and successfully applied scheme, a thin wire is placed behind the phase modulating element [114]. The opaque wire is the origin of a (near-) cylindrical diffracted wave that interferes with the original (TC = 1) Vortex beam. At some point along the wire, the phase discontinuity will become noticeable in that the interference term changes sign. Accordingly, a fork in the fringe pattern, similar to the one in Fig. 5.2, indicates the π phase jump between opposing parts of the wave front (recalling Fig. 5.1 **b**, for each angular direction away from the beam centre, the phase across the singularity is removed by a value of π). For higher charges, the fork will be manifold, equal to the TC.

We simulated this pattern for the approximate conditions that would apply in our case, showing the result in Fig. 6.5 **a**. In ref. [114], a pre-existing x-ray beam is modified to carry OAM. We wanted to implement the wave front splitting scheme into our experiment, where the Vortex phase feature is imprinted on a fundamental, but the light field under investigation is only generated much later, in an High-Harmonic setup. This meant for our case that we had to place the wire shortly behind the gas target, utilising the beamline and detection method as seen in Fig. C.1 **b**. The simulation includes an idealised version of our spatial profile—i.e., one with a full circle around a dark Vortex core. The vertically placed wire is fully opaque, and we looked at the diffraction pattern after a propagation distance of two Rayleigh lengths. As expected, following a bright interference fringe, once we move over the singularity, the interference switches to destructive, and we hit a dark fringe. We have added dashed white lines to guide the eye.

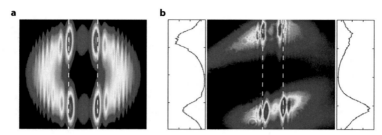

Figure 6.5: Simulation (**a**) and experiment (**b**) of the investigation of the phase of an XUV Optical Vortex by division of the wave front. In each sub-figure, dashed lines guide the eye along the 'ridge' of one bright interference fringe. On the opposing side of the beam, the extension of the line coincides with a dark fringe. Greater TCs would result in shifts by more than one fringe, which would be visible as a swirling distortion of the profiles. In **b**, we also show measured profiles of the signal along the indicated lines, revealing the asymmetry in fringe brightness (Image **a** courtesy of A. Dreischuh).

In our experimental case, the best achievable spatial modes consist, as was shown, of two separate lobes of intensity. Under ideal conditions, signals from both lobes are nearly equal. In contrast to experiments presented in the previous section, we use the full laser power for HHG, which suggests that several modes of the harmonic comb contribute to the XUV signal. Under these regular conditions, the majority of the signal is produced by the dominant 23rd harmonic, at $\sim 35\,\text{nm}$—the previously discussed simulation was performed at the same (monochromatic) wavelength. We placed a tungsten wire ($\varnothing \approx 5\,\mu\text{m}$) a few centimetres behind the nickel tube and viewed the resulting spatial profile in the far-field behind the wire. We present the image in Fig. 6.5 **b**. The wire is also oriented vertically in this image. Although the signal is interrupted horizontally, as before, we can trace the occurring interference fringes. Again, we guide the eye with dashed lines that are drawn by aiming along the 'ridge' of a bright fringe. We clearly observe that each line hits a dark fringe exactly when we follow its respective path. In spite of poor fringe contrast, the asymmetry is visible, confirmed by two profiles that trace the signal along the closest dashed line. This fits very well with the prediction from the simulation, indicating that the phase behaviour of our HHG light is described by an Optical Vortex.

We encountered a peculiar point here that is observed throughout all of our investigations, namely that the best agreement of the evaluated features and the expected or simulated behaviour occurs under the assumption that the emerging OV in the XUV only carries a single unit of Topological Charge. Simulations, even for low-order TC greater than unity, show that the swirling distortion of the fringe pattern is enhanced with increasing TC. This is not surprising since, for each integer of TC, contrast of one more multiple of 2π

has to be accommodated into opposing fringes. The experimentally recorded image in Fig. 6.5 **b** shows no indication of such multiple forking signatures. We shall address the apparent exclusiveness of singular charged Vortices for our conditions in later sections of this chapter, after some more dedicated experimental investigation.

Phase evaluation by a Hartmann-type setup

Another concept that is commonly used in the evaluation of phase profiles of beams is known as the (Shack-) Hartmann wave front sensor. This is a purely geometrical-optic technique that samples an impinging beam profile at a certain number of points. The pure Hartmann-type of this technique does this by allowing single sections of a beam to pass through holes in an otherwise opaque mask. The holes are much larger than the wavelength, so that an undiffracted clear-cut ray of light is issued from each hole. The direction in which this ray is inclined in the space behind the mask, falling onto a detector, is determined by the local tilt of the phase front, or position-specific Poynting vector, in the original beam. In reference to a plane wave for calibration purposes, the resulting shift of the centre-of-mass of each point in the resulting spot pattern on the detector can be globally summed up and used to reconstruct the shape of the complete wave front. This happens at a sampling resolution that is determined by the fabrication of the mask, and the distance and resolution of the detector. Shack's extension to this scheme later merely added tiny lenses at each sampling point, allowing a more accurate evaluation of the shifts by looking at focal spots on the detector.

We used a home-built Hartmann device for measuring aberrations in High-Harmonic beams. More details on the working principle of the device, including all parameters and information on the retrieval algorithms, can be found in [140, 145]. Applications of the concept on our High-Harmonic source is found in [18, 19]. It should suffice to say here that the applied mask is inserted at a well-defined distance in front of the same large area CCD as shown in Fig. C.1 **b**. The mask comprises a rectangular array of 50×50 holes of $40 \, \mu m$ diameter and $140 \, \mu m$ separation. This allows the covering of a beam area of $\sim 7 \times 7 \, mm^2$, at a phase resolution of $\lesssim 10 \, nm$.

With such a setup, it is always possible to extract the local phase front tilt information, because each single point is initially evaluated at its exact position in reference to the point projection from a plane wave. This is true regardless of the overall shape of the wave front, including OV shapes as researched here. However, in the more special Vortex scenarios, the conventionally applied summing-up-algorithms for reconstruction of the full phase front fail due to the phase ambiguity. More precisely, it becomes dependent upon the (discrete numerical) integration path that is chosen to cover information from *all* points illuminated by the beam. The continuous rotational symmetry of the phase prohibits conventional algorithms from finding the phase branch and the ensuing 2π jump illustrated in Fig. 5.1 **b**. However, certain iterative procedures that narrow down the search for the point singularity using looping paths, have been proposed by Fried [146], and successfully applied in very well-defined Vortex beams [147, 148].

However, as should be clear from the concept of the Hartmann sensor described above, the whole method relies on the presence of actual signal from the spots. This is already undermined by the missing intensity in the centre of a typical Vortex beam. A dark Hartmann spot does not necessarily form a substantial problem, as the centre-of-mass of small signals will still be evaluated. Due to noise, this will result in a statistically placed point of support for reconstruction. In the good quality beams from the above cited examples, phase front evaluation was still possible. This is, however, far from the quality of our obtained fragmented intensity profiles that have already been shown in Figs. 6.3 and 6.4. Also, the area of coverage of our available Hartmann mask was only large enough to sample the majority of the Vortex beam, while neglecting outer regions of intensity. We addressed these issues by only considering sections of the wave front that were sampled at a significant signal level, and by only summing up the relative spot shifts within these limited regions. The results are presented in Fig. 6.6.

Shown in sub-figure **a** is an intensity plot of the Vortex beam as it covers the Hartmann area, at an arbitrary false-colour scale. It was produced under identical conditions as the beam that was evaluated by the division-of-wave-front setup in the preceding paragraph. The intensity value is obtained from the fully integrated signal sampled at each Hartmann hole. One can clearly see the two near-equal lobes in the upper-left and lower-right corners of the mask, and the large dark area in between. This illustrates the portion of the beam that could be evaluated by this setup. In sub-figure **b**, we show the detected wave front shifts, where the advancement or retardation of the phase in propagation (z-) direction has been calculated from the summation of x- and y-components of the centre-of-mass-shifts of the detected spots. The summation paths that lead to the reconstruction of the observed segments in the figure do not include or encircle the singularity. Thus, we assume that no problematic ambiguities compromise the measurement as presented here.

As a reference, we used the spot pattern of the fully illuminated mask, taken under the exact same HHG conditions, only with a flat phase applied to the SLM. The Gaussian profile should then produce a reference spot pattern that includes the divergence and intrinsic deviations of the beam. These are thus already considered and corrected for

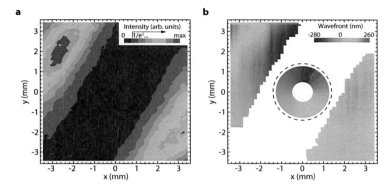

Figure 6.6: Phase front measurement of an Optical Vortex in the XUV by a Hartmann-type setup. **a**, Intensity profile of the part of the OV beam that could be evaluated by our Hartmann mask. **b**, Only the data from Hartmann holes with intensities greater than $1/e^2$ of the maximum signal are plotted here. Shown is the advancement or retardation of points of equal phase in propagation direction, calculated from the mean centre-of-mass shift between neighbouring holes. The shifts are summed up along 'allowed' paths, sparing the dark core. An arbitrarily scaled spiral in the centre of the sub-figure is added to display the expected angular distribution of phase height in the case of a perfect OV beam for comparison.

by the reference itself. From the intensity profile in Fig. 6.6 **a**, we considered only the points at which the signal of the Vortex beam was above the $1/e^2$ level of the maximum, resulting in the plotted sections in **b**. For comparison we also show, in the centre of the plot, an arbitrarily scaled helical profile, with an orientation of the branch point at an angle that shows best accordance with the colour scale that emerged from plotting the measured phase front.

It should be stressed that we want to highlight here only the qualitative features of the local slopes in the phase front that fit very well to expectations from a helical profile. The absolute quantitative scale, as shown in Fig. 6.6 **b**, would correspond to more than 10 wavelengths, assuming, as before, that the $23^{\rm rd}$ harmonic is dominant. This calibration is subject to a number of factors, and should be taken with a grain of salt. We shall not attempt to make any statements about the number of windings, or TC, without more reliable numbers. However, we take the evidence of counter-directional tilt of the phase front in the two opposing lobes as an indicator for a helical feature in the XUV beam generated with an OV in the fundamental.

This concludes the investigations of the phase of the claimed XUV Vortex. What is still undetermined is the behaviour of the generated Vortex beams in different orders of HHG.

For this, we combined wavelength resolution using a spectrometer with the investigation of spatial features of our beam. The already mentioned peculiarity of the apparent single unit of Topological Charge that was deduced from the pattern in Fig. 6.5 **b** shall be under special consideration in the following experiments.

6.3.3 XUV Vortex beams in different harmonic orders

The extent of the spatial intensity feature of an OV has been introduced to be dependent upon the Topological Charge of the Vortex in question—the higher the charge, the larger the dark core becomes. As the full electric field, including phase, is multiplied by the harmonic order in a nonlinear process, one should also expect to observe higher Vortex charges in higher harmonic numbers. We shall discuss the expected behaviour, in the light of a qualitative argument and examples mentioned in section 5.4, in the upcoming results section of this part. But first, let us describe the methods and findings of our evaluation.

For this line of investigation, we employed the beamline as described in Fig. C.1 **c**. HHG conditions were the same as in the previous section, in that we used the full, unattenuated laser output. This means that, regardless of the phase shape modulated onto the fundamental, the resulting beam comprises a number of harmonics around 30–50 nm. As the focussed mode size of the fundamental drastically grows with Topological Charge (see Fig. 6.2), the cut-off decreases due to lower intensities. We restrict ourselves to examine only Vortex orders of 1 and 2, as only these produce evaluable signal.

Initially, the spectrometer was used with a wide opened entrance slit. The grating was then illuminated by a non-collimated beam. Its diffractive effect, with appropriate choice of line density, still works to separate harmonics of different wavelengths. However, the angular spread of the beam at the site of the slit overlays this energy selection in the dispersion sensitive direction with a (slightly distorted) spatial image of the evaluated beam. Perpendicular to that, all spatial features can be recorded by the imaging CCD in the spectrometer without additional spectral distortion. The result for a non-collimated Vortex beam is shown in Fig. 6.7 **a**. We see that the two-lobed feature of the hitherto only spatially resolved beam is self-similarly imprinted onto all harmonic orders, as far as sufficient signal is available.

This is the first indication that the premises formulated at the beginning of this section are questioned by our experiment. To further quantify our findings, we looked more closely on the accurately reproduced spatial features in the vertical pixel direction of the imaging spectrometer. For this, we closed the slit to gain a nearly collimated beam, allowing the brightest regions in each harmonic order to pass through for spectral evaluation. With

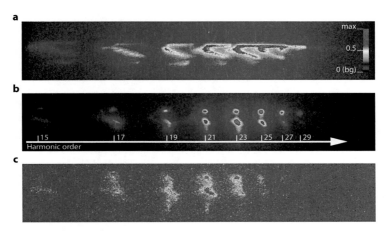

Figure 6.7: Spectral evidence of Vortex signature in all visible harmonic orders. **a**, OV with TC = 1 applied. With an opened entrance slit of the spectrometer, we gain spectral separation of the harmonic orders, but also retain some spatial information about the beam profile vertically. All orders look similar, as in Fig. 6.3. **b**, Same as **a**, but with the entrance slit closed to only allow the most intense regions of the beam to hit the spectrometer grating. Spatial information in the vertical dimension is presented without distortion, as it is unaffected by the grating. **c**, Same as **b**, but with a TC of 2 applied to the SLM. All images are presented in the shown false-colour scale, however with respective normalisation.

some trade-off in accuracy due to slight asymmetries in different harmonic orders, this was achieved for applied TCs of 1 and 2. The respective results are shown in Figs. 6.7 **b** and **c**. Now, for each significant harmonic order, and applied TC, we selected an horizontal pixel window that incorporated all signal in each order, with the full height of the CCD window. We numerically integrated the signal over the width of the window, and plotted intensity along the spatially resolved vertical chip dimension. The results form the (black) signal plots of Fig. C.2 in appendix C.

For a certain number of harmonics generated with a fundamental carrying a TC of 1 (Fig. C.2 **b**), and a somewhat smaller number generated with a TC of 2 (**c**), the two separate lobes of signal are clearly observed. We use a phenomenological fit of the signal (Fig. C.2, red lines) as a sum of single Lorentzian peaks (green lines) to gain a quantitative description of the lobe distance. On a pixel scale, the distance is noted in each subplot. It is then converted to an actual length scale via pixel size and plotted, with errors stemming from the fits, against the measured wavelength of each harmonic order in Fig. 6.8 **a**. For convenience, we use wavelengths as a spectral unit, as it is used throughout this work, but plot increasing energy or harmonic order (and *de*creasing wavelength) to the right.

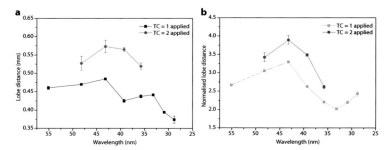

Figure 6.8: Vortex-core size in different harmonic orders, as gained from the pixel distance of peaks in Fig. C.2. **a**, For TC = 1 and 2, the pixel distance has been converted to a length and plotted vs. the measured wavelength of the harmonic. Error bars stem from errors of the fit parameters in Fig. C.2. **b**, We take the same data as for sub-figure **a**, but normalise for the full-width at half-maximum of the HHG signal in respective harmonic order, generated with a Gaussian. By doing this, we compensate for intrinsic divergence in each order, which differs slightly due to phase matching.

What we see is that there is no clear trend in the size of the spatial Vortex signature generated with a specific TC in the fundamental. In particular, there is no proportional increase with harmonic order, as was implied at the outset of this investigation. However, we do observe a significant increase in Vortex size once we imprint a TC of 2 onto the fundamental, over the 'TC = 1' case. This feature persists when we compensate for the intrinsic divergence of the High-Harmonic signal in different orders, due to specific phase matching conditions in each case. We do this by normalising the peak distance with the FWHM width of a Gaussian fit on the harmonics obtained from an experiment with a flat phase applied to the SLM. The fits and obtained parameters are shown in Fig. C.2 **a** (red lines). The resulting normalised lobe distances are plotted against wavelength in Fig. 6.8 **b**. No clear trend is visible in this presentation either, but the greater feature size of 'TC = 1'- over 'TC = 2'-generated harmonics follows the same line in each case.

These findings imply that the Topological Charge, or degree of vorticity, does not simply multiply with harmonic order in the experiment described here. Within a certain experimental tolerance, we might even say that the same TC is retained within each generation scenario, whereas a higher fundamental TC carries over to a greater Vortex size, or a minimal obtained TC, in the harmonics. We offer some explanations, and make a few comparisons in the context of recent findings of other groups, in the following chapter.

Chapter 7

Summary and Discussion (part II)

This concludes the presentation of our experimental findings on the Generation of High-Harmonics with a vorticity imprinted on the fundamental driving laser. Spatial and phase-sensitive investigations strongly indicate that we have indeed transplanted signatures of an Optical Vortex over a non-perturbative high-order process onto light in the extreme ultraviolet. This works simultaneously for a number of harmonic modes within a specific phase-matched spectral envelope, between the 15th and 29th order.

Spatially, we observe reproducible intensity profiles incorporating the expected dark Vortex core at a position that is defined as the central symmetric axis by 'conventional' Gaussian HHG. We did not observe a full-circle OV profile as would be described by a well-defined Laguerre-Gaussian mode. Instead, we report separated lobes of near equal intensity that are formed in opposing sectors around the dark centre. Considering the inadequacies of our fundamental beam profile, generating such an imperfect structure seems plausible, as a high degree of nonlinearity would be expected to enhance such features. We can, however, safely assume that the formation of these lobes is not just due to spatially varying phase matching away from the optical axis. When we changed gas pressure—a typical handle on phase matching conditions—the geometry of the Vortex beam, including location and size of the dark core, remained completely unchanged over the full accessible parameter range.

Adaptation of a Hartmann-type phase evaluation reveals that, in reference to a Gaussian HHG beam under otherwise identical conditions, the spatial phase is sloped in opposite directions inside the opposing lobes. This is exactly what one would expect if the phase front describes a helical profile. We also examined phase behaviour using a division-of-wave-front method and observed a half-period (bright-dark) fringe shift in phase between the lobes. Judging by all of the information gained in that experiment, the best match between simulated and measured interference patterns suggests a single winding over the full circle of the beam, which would imply a Topological Charge of one in our Vortex HHG beam.

An intuitive expectation, in that the TC of the fundamental Vortex should be multiplied by the order of the process to produce highly charged OVs in the XUV, could not be documented. We conclude this from the interferometric observation mentioned above, but more specifically from the evaluation of the Vortex core size—or spatial distance of the two signal maxima—under consideration of the harmonic order in which the Vortex is generated. The multiplicity of the phase in the Mth harmonic is, however, an intrinsic feature that can be gleaned from the basic properties of the Fourier transform of a periodic process, as was, in fact, formulated in Eq. (2.17) in this work. A soliton model of the propagation of highly charged OVs has found strong arguments that a modulational instability leads to a break up of Vortices into 'daughter' solitons of smaller, and finally fundamental TC, conserving the total number of angular momentum quanta [138, 139]. Furthermore, it has been postulated and observed that TCs of equal sign co-rotate in propagation, and repel each other in a nonlinearly behaving background [149, 150]. Thus, it seems plausible to assume that high-TC Vortices that are generated inside the target gas split and diverge, while only fundamental OVs arrive at the distant detector. Still, this does not account for the fact that the apparent fundamental TC should be different depending on whether the driving field carries a TC of 1 or 2—the fundamental charge of *any* degree of vorticity should be one. We see, however, a clear difference in the spatial modes generated using different fundamental TC, as documented in Figs. 6.7–6.8.

A recent publication by Hernández-García *et al.* numerically investigates the properties of HHG with an OV in the driving field, under similar conditions as we used experimentally [151]. They simulate realistic propagation of the harmonics, however in a parameter regime where instabilities due to nonlinearity can be neglected. Once the High-Harmonics are emitted by discrete, phase matched radiators inside the Gaussian-density-profiled interaction region, further propagation is assumed strictly linear. Under these conditions, the authors confirm the expected multiplicity of the TC by the harmonic number. In the meantime, Corkum *et al.* have succeeded in verifying high-TC Vortices of expected order in an ingenious setup allowing interferometry of single harmonics [152]. They also state that, for shorter wavelengths than that of the 15th harmonic which they researched, aberrations and defects in the setup may be detrimental to the experiment.

In [151], simulation of the full phase profile shows clearly that large TCs are produced in the respective harmonic order (e.g. TC = 17 in the 17th harmonic), with a fundamental $LG_{1,0}$ mode. The spatial cross-sections, on the other hand, are generated at similar core sizes, as seen in Fig. 7.1. Notably, the harmonics produced from an $LG_{2,0}$ fundamental exhibit a slightly greater divergence than the $LG_{1,0}$ harmonics, although they are nearly

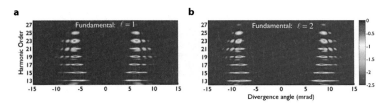

Figure 7.1: Simulated spatial divergence of High-Harmonic signal generated with Topological Charges in the fundamental of 1 (**a**) and 2 (**b**) (from [151]).

identical in size when compared to one another. In our experiments we only evaluated sections of the two-lobed intensity profiles, as was described. Now, the simulations show that the expectation that Vortex core sizes just visibly increases seems unjustified. Further experimental investigations conducted along those lines are recommended to improve the spatial profile of the generated OVs. It would be a first instructive step to verify whether an optimised fundamental beam profile would succeed in producing a full-circle OV signature. With such a beam, a cleaner interference pattern for a measurement as shown in Fig. 6.5 might be obtained. It should be attempted to perform this measurement also with spectral resolution. With more data like this, a definite answer might be found as to whether decay of the TC due to an instability is present, or if we simply lacked sufficient resolution or signal to observe the actual phase difference in opposing lobes.

Fortunately, as was also verified in the simulations in [151], the temporal structure of helical HHG benefits greatly from the near-identical divergence of the harmonics in the comb. For purposes of attosecond experiments, the full field is necessary in order to reach the shortest pulse durations. It would be detrimental for such a scheme if the harmonics in greater orders were separated by the divergence due to greater Vortex charge. Hernández-García *et al.* go so far as to propose that, using a few-cycle driving field, different settings of the carrier-envelope phase occur almost simultaneously at different angular orientations within the same beam. This might enable an impressive additional degree of freedom in experiments as the ones we cited as a driving force behind our investigations (see section 5.3 of this work and refs. [128, 129]).

There remain many challenges in testing the theory and fundamental concepts forming the basis of this work, as well as making XUV Vortices with coherence properties of High-Harmonics into a regular tool in applications that benefit from helical light. What we have learnt from our experiments is encouraging to further pursue this line of investigation.

Part III

Literature, Lists and Appendices

Bibliography

[1] T. Pfeifer, C. Spielmann, and G. Gerber. Femtosecond x-ray science. *Rep. Prog. Phys.*, 69(2):443–505, 2006.

[2] G. Sansone, L. Poletto, and M. Nisoli. High-energy attosecond light sources. *Nat Photon*, 5(11):655–663, 2011.

[3] P. B. Corkum and F. Krausz. Attosecond science. *Nat Phys*, 3(6):381–387, 2007.

[4] P. B. Corkum. Plasma perspective on strong field multiphoton ionization. *Phys. Rev. Lett.*, 71(13):1994, 1993.

[5] J. L. Krause, K. J. Schafer, and K. C. Kulander. High-order harmonic generation from atoms and ions in the high intensity regime. *Phys. Rev. Lett.*, 68(24):3535–3538, 1992.

[6] M. Ferray, A. L'Huillier, X. F. Li, L. A. Lompré, G. Mainfray, and C. Manus. Multiple-harmonic conversion of 1064 nm radiation in rare gases. *Journal of Physics B: Atomic, Molecular and Optical Physics*, 21(3):L31, 1988.

[7] T. Ditmire, E. T. Gumbrell, R. A. Smith, J. W. G. Tisch, D. D. Meyerhofer, and M. H. R. Hutchinson. Spatial Coherence Measurement of Soft X-Ray Radiation Produced by High Order Harmonic Generation. *Phys. Rev. Lett.*, 77(23):4756–4759, 1996.

[8] T. Brabec and F. Krausz. Intense few-cycle laser fields: Frontiers of nonlinear optics. *Rev. Mod. Phys.*, 72(2):545, 2000.

[9] E. Goulielmakis, M. Schultze, M. Hofstetter, V. S. Yakovlev, J. Gagnon, M. Uiberacker, A. L. Aquila, E. M. Gullikson, D. T. Attwood, R. Kienberger, F. Krausz, and U. Kleineberg. Single-Cycle Nonlinear Optics. *Science*, 320(5883):1614–1617, 2008.

[10] T. Popmintchev, M. C Chen, P. Arpin, M. M. Murnane, and H. C. Kapteyn. The attosecond nonlinear optics of bright coherent X-ray generation. *Nat Photon*, 4(12):822–832, 2010.

[11] F. Krausz and M. Y. Ivanov. Attosecond physics. *Rev. Mod. Phys.*, 81(1):163–234, 2009.

[12] H. J. Wörner, J. B. Bertrand, B. Fabre, J. Higuet, H. Ruf, A. Dubrouil, S. Patchkovskii, M. Spanner, Y. Mairesse, V. Blanchet, E. Mével, E. Constant, P. B. Corkum, and D. M. Villeneuve. Conical Intersection Dynamics in NO2 Probed by Homodyne High-Harmonic Spectroscopy. *Science*, 334(6053):208–212, 2011.

[13] A. N. Pfeiffer, C. Cirelli, M. Smolarski, and U. Keller. Recent attoclock measurements of strong field ionization. *Attosecond spectroscopy*, 414(0):84–91, 2013.

[14] E. Takahashi, Y. Nabekawa, T. Otsuka, M. Obara, and K. Midorikawa. Generation of highly coherent submicrojoule soft x rays by high-order harmonics. *Phys. Rev. A*, 66(2):021802, 2002.

[15] A. D. Shiner, C. A. Trallero-Herrero, N. Kajumba, H. C Bandulet, D. Comtois, F. Légaré, M. Giguère, J.-C. Kieffer, P. B. Corkum, and D. M. Villeneuve. Wavelength Scaling of High Harmonic Generation Efficiency. *Phys. Rev. Lett.*, 103(7):073902, 2009.

[16] P. Zeitoun, P. Balcou, S. Bucourt, F. Delmotte, G. Dovillaire, D. Douillet, J. Dunn, G. Faivre, M. Fajardo, K. A. Goldberg, S. Hubert, J. R. Hunter, M. Idir, S. Jacquemot, S. Kazamias, S. Le Pape, X. Levecq, C. L. S. Lewis, R. Marmoret, P. Mercère, A.-S Morlens, P. P. Naulleau, M. F. Ravet, C. Rémond, J. J. Rocca, R. F. Smith, P. Troussel, C. Valentin, and L. Vanbostal. Recent developments in X-UV optics and X-UV diagnostics. *Appl. Phys. B*, 78(7-8):983–988, 2004.

[17] J. Gautier, P. Zeitoun, C. Hauri, A.-S Morlens, G. Rey, C. Valentin, E. Papalarazou, J.-P Goddet, S. Sebban, F. Burgy, P. Mercère, M. Idir, G. Dovillaire, X. Levecq, S. Bucourt, M. Fajardo, H. Merdji, and J.-P Caumes. Optimization of the wave front of high order harmonics. *The European Physical Journal D*, 48(3):459–463, 2008.

[18] J. Lohbreier, S. Eyring, R. Spitzenpfeil, C. Kern, M. Weger, and C. Spielmann. Maximizing the brilliance of high-order harmonics in a gas jet. *New J. Phys.*, 11(2):023016, 2009.

[19] R. Spitzenpfeil, S. Eyring, C. Kern, C. Ott, J. Lohbreier, J. Henneberger, N. Franke, S. Jung, D. Walter, M. Weger, C. Winterfeldt, T. Pfeifer, and C. Spielmann.

Enhancing the brilliance of high-harmonic generation. *Appl. Phys. A*, 96(1):69–81, 2009.

[20] S. Eyring, C. Kern, M. Zürch, and C. Spielmann. Improving high-order harmonic yield using wavefront-controlled ultrashort laser pulses. *Opt. Express*, 20(5):5601–5606, 2012.

[21] K. Nakajima. Compact X-ray sources: Towards a table-top free-electron laser. *Nat Phys*, 4(2):92–93, 2008.

[22] R. J. Jones, K. D. Moll, M. J. Thorpe, and J. Ye. Phase-Coherent Frequency Combs in the Vacuum Ultraviolet via High-Harmonic Generation inside a Femtosecond Enhancement Cavity. *Phys. Rev. Lett.*, 94(19):193201, 2005.

[23] M. C Chen, M. R. Gerrity, S. Backus, T. Popmintchev, X. Zhou, P. Arpin, X. Zhang, H. C. Kapteyn, and M. M. Murnane. Spatially coherent, phase matched, high-order harmonic EUV beams at 50 kHz. *Opt. Express*, 17(20):17376–17383, 2009.

[24] S. Hädrich, J. Rothhardt, M. Krebs, F. Tavella, A. Willner, J. Limpert, and A. Tünnermann. High harmonic generation by novel fiber amplifier based sources. *Opt. Express*, 18(19):20242–20250, 2010.

[25] M. Krebs, S. Hädrich, S. Demmler, J. Rothhardt, A. Zaïr, L. Chipperfield, J. Limpert, and A. Tünnermann. Towards isolated attosecond pulses at megahertz repetition rates. *Nat Photon*, 7(7):555–559, 2013.

[26] A. Vernaleken, J. Weitenberg, T. Sartorius, P. Russbueldt, W. Schneider, S. L. Stebbings, M. F. Kling, P. Hommelhoff, H.-D. Hoffmann, R. Poprawe, F. Krausz, T. W. Hänsch, and T. Udem. Single-pass high-harmonic generation at 20.8 MHz repetition rate. *Opt. Lett.*, 36(17):3428–3430, 2011.

[27] E. Seres, J. Seres, and C. Spielmann. Extreme ultraviolet light source based on intracavity high harmonic generation in a mode locked Ti:sapphire oscillator with 9.4 MHz repetition rate. *Opt. Express*, 20(6):6185–6190, 2012.

[28] T. Pfeifer, R. Spitzenpfeil, D. Walter, C. Winterfeldt, F. Dimler, G. Gerber, and C. Spielmann. Towards optimal control with shaped soft-x-ray light. *Opt. Express*, 15(6):3409–3416, 2007.

[29] S. Kim, J. Jin, Y.-J. Kim, I.-Y. Park, Y. Kim, and S.-W. Kim. High-harmonic generation by resonant plasmon field enhancement. *Nature*, 453(7196):757–760, 2008.

[30] W. L. Barnes, A. Dereux, and T. W. Ebbesen. Surface plasmon subwavelength optics. *Nature*, 424(6950):824–830, 2003.

[31] M. Kauranen and A. V. Zayats. Nonlinear plasmonics. *Nat Photon*, 6(11):737–748, 2012.

[32] R. Petry, M. Schmitt, and J. Popp. Raman Spectroscopy—A Prospective Tool in the Life Sciences. *Chem. Phys. Chem.*, 4(1):14–30, 2003.

[33] G. Herink, D. R. Solli, M. Gulde, and C. Ropers. Field-driven photoemission from nanostructures quenches the quiver motion. *Nature*, 483(7388):190–193, 2012.

[34] M. Krüger, M. Schenk, M. Förster, and P. Hommelhoff. Attosecond physics in photoemission from a metal nanotip. *Journal of Physics B: Atomic, Molecular and Optical Physics*, 45(7):074006, 2012.

[35] A. Husakou, S.-J. Im, and J. Herrmann. Theory of plasmon-enhanced high-order harmonic generation in the vicinity of metal nanostructures in noble gases. *Phys. Rev. A*, 83(4):043839, 2011.

[36] M. F. Ciappina, T. Shaaran, and M. Lewenstein. High order harmonic generation in noble gases using plasmonic field enhancement. *Annalen der Physik*, 525(1-2):97–106, 2013.

[37] A. Husakou, F. Kelkensberg, J. Herrmann, and M. J. J. Vrakking. Polarization gating and circularly-polarized high harmonic generation using plasmonic enhancement in metal nanostructures. *Opt. Express*, 19(25):25346–25354, 2011.

[38] N. Pfullmann. *Nano-antenna-assisted high-order harmonic generation*. Dissertation, Gottfried Wilhelm Leibniz Universität, Hannover, 2012.

[39] M. Sivis, M. Duwe, B. Abel, and C. Ropers. Nanostructure-enhanced atomic line emission. *Nature*, 485(7397):E1–E2, 2012.

[40] M. Sivis, M. Duwe, B. Abel, and C. Ropers. Extreme-ultraviolet light generation in plasmonic nanostructures. *Nat Phys*, 9(5):304–309, 2013.

[41] I.-Y. Park, J. Choi, D.-H. Lee, S. Han, S. Kim, and S.-W. Kim. Generation of EUV radiation by plasmonic field enhancement using nano-structured bowties and funnel-waveguides. *Annalen der Physik*, 525(1-2):87–96, 2013.

[42] L. V. Keldysh. Ionization in Field of a Strong Electromagnetic Wave. *Soviet Physics JETP-USSR*, 20(5):1307–, 1965.

[43] M. V. Ammosov, N. B. Delone, and V. P. Krainov. Tunnel Ionization of Complex Atoms and Atomic Ions in a Varying Electromagnetic-Field. *Zh. Eksp. Teor. Fiz.*, 91(6):2008–2013, 1986.

[44] J. Mauritsson, P. Johnsson, E. Gustafsson, A. L'Huillier, K. J. Schafer, and M. B. Gaarde. Attosecond Pulse Trains Generated Using Two Color Laser Fields. *Phys. Rev. Lett.*, 97(1):013001, 2006.

[45] M. Lewenstein, P. Balcou, M. Y. Ivanov, A. L'Huillier, and P. B. Corkum. Theory of high-harmonic generation by low-frequency laser fields. *Phys. Rev. A*, 49(3):2117, 1994.

[46] M. Lewenstein, P. Salières, and A. L'Huillier. Phase of the atomic polarization in high-order harmonic generation. *Phys. Rev. A*, 52(6):4747, 1995.

[47] R. W. Boyd. *Nonlinear Optics, Third Edition*. Academic Press, 2008.

[48] P. Balcou, P. Salières, A. L'Huillier, and M. Lewenstein. Generalized phase-matching conditions for high harmonics: The role of field-gradient forces. *Phys. Rev. A*, 55(4):3204–3210, 1997.

[49] M. B. Gaarde, F. Salin, E. Constant, P. Balcou, K. J. Schafer, K. C. Kulander, and A. L'Huillier. Spatiotemporal separation of high harmonic radiation into two quantum path components. *Phys. Rev. A*, 59(2):1367, 1999.

[50] T. Auguste, P. Salières, A. S. Wyatt, A. Monmayrant, I. A. Walmsley, E. Cormier, A. Zaïr, M. Holler, A. Guandalini, F. Schapper, J. Biegert, L. Gallmann, and U. Keller. Theoretical and experimental analysis of quantum path interferences in high-order harmonic generation. *Phys. Rev. A*, 80(3):033817, 2009.

[51] P. Muhlschlegel, H.-J. Eisler, O. J. F. Martin, B. Hecht, and D. W. Pohl. Resonant optical antennas. *Science*, 308(5728):1607–1609, 2005.

[52] L. Novotny and N. van Hulst. Antennas for light. *Nat Photon*, 5(2):83–90, 2011.

[53] J. N. Anker, W. P. Hall, O. Lyandres, N. C. Shah, J. Zhao, and R. P. van Duyne. Biosensing with plasmonic nanosensors. *Nat Mater*, 7(6):442–453, 2008.

[54] L. Novotny and S. J. Stranick. Near-Field Optical Microscopy and Spectroscopy with Pointed Probes. *Annu. Rev. Phys. Chem*, 57(1):303–331, 2006.

[55] S. Pillai, K. R. Catchpole, T. Trupke, and M. A. Green. Surface plasmon enhanced silicon solar cells. *J. Appl. Phys.*, 101(9):093105, 2007.

[56] J. Jahns and S. Helfert. *Introduction to micro- and nanooptics*. Wiley-VCH-Verl., Weinheim, 2012.

[57] L. Novotny and B. Hecht. *Principles of Nano-optics*. Cambridge University Press, Cambridge, second edition, 2012.

[58] M. I. Stockman. Nanoplasmonics: past, present, and glimpse into future. *Opt. Express*, 19(22):22029–22106, 2011.

[59] P. B. Johnson and R. W. Christy. Optical Constants of the Noble Metals. *Phys. Rev. B*, 6(12):4370–4379, 1972.

[60] G. Mie. Beiträge zur Optik trüber Medien, speziell kolloidaler Metallösungen. *Annalen der Physik*, 330(3):377–445, 1908.

[61] G. Raschke. *Molekulare Erkennung mit einzelnen Gold–Nanopartikeln*. Dissertation, Ludwig–Maximilians–Universität, München, 2005.

[62] S. Link and M. A. El-Sayed. Optical properties and ultrafast dynamics of metallic nanocrystals. *Annu. Rev. Phys. Chem*, 54:331–366, 2003.

[63] M. S. Tame, K. R. McEnery, S. K. Ozdemir, J. Lee, S. A. Maier, and M. S. Kim. Quantum plasmonics. *Nat Phys*, 9(6):329–340, 2013.

[64] J. D. Jackson. *Classical electrodynamics*. Wiley, 1975.

[65] B. T. Draine and P. J. Flatau. Discrete-dipole approximation for scattering calculations. *J. Opt. Soc. Am. A*, 11(4):1491–1499, 1994.

[66] A. Taflove and S. C. Hagness. *Computational Electrodynamics: The Finite-Difference Time-Domain Method*. Artech House, Boston, third edition, 2010.

[67] C. Rockstuhl, T. Zentgraf, H. Guo, N. Liu, C. Etrich, I. Loa, K. Syassen, J. Kuhl, F. Lederer, and H. Giessen. Resonances of split-ring resonator metamaterials in the near infrared. *Appl. Phys. B*, 84(1-2):219–227, 2006.

[68] L. Novotny. Effective wavelength scaling for optical antennas. *Phys. Rev. Lett.*, 98(26), 2007.

[69] K. C. Y. Huang, Y. C. Jun, M.-K. Seo, and M. L. Brongersma. Power flow from a dipole emitter near an optical antenna. *Opt. Express*, 19(20):19084–19092, 2011.

[70] E. Cubukcu, N. Yu, E. J. Smythe, L. Diehl, K. B. Crozier, and F. Capasso. Plasmonic Laser Antennas and Related Devices. *IEEE Journal of Selected Topics in Quantum Electronics*, 14(6):1448–1461, 2008.

[71] R. D. Grober, R. J. Schoelkopf, and D. E. Prober. Optical antenna: Towards a unity efficiency near-field optical probe. *Appl. Phys. Lett*, 70(11):1354–1356, 1997.

[72] K. B. Crozier, A. Sundaramurthy, G. S. Kino, and C. F. Quate. Optical antennas: Resonators for local field enhancement. *J. Appl. Phys.*, 94(7):4632–4642, 2003.

[73] R. Marty, G. Baffou, A. Arbouet, C. Girard, and R. Quidant. Charge distribution induced inside complex plasmonic nanoparticles. *Opt. Express*, 18(3):3035–3044, 2010.

[74] J. Merlein. *Lineare und nichtlineare Nanoplasmonik*. Dissertation, Universität Konstanz, 2008.

[75] D. P. Fromm, A. Sundaramurthy, P. J. Schuck, G. S. Kino, and W. E. Moerner. Gap-Dependent Optical Coupling of Single "Bowtie" Nanoantennas Resonant in the Visible. *Nano Lett*, 4(5):957–961, 2004.

[76] P. J. Schuck, D. P. Fromm, A. Sundaramurthy, G. S. Kino, and W. E. Moerner. Improving the Mismatch between Light and Nanoscale Objects with Gold Bowtie Nanoantennas. *Phys. Rev. Lett.*, 94(1):017402, 2005.

[77] H. Guo, T. P. Meyrath, T. Zentgraf, N. Liu, L. Fu, H. Schweizer, and H. Giessen. Optical resonances of bowtie slot antennas and their geometry and material dependence. *Opt. Express*, 16(11):7756–7766, 2008.

[78] S. Park, J. W. Hahn, and J. Y. Lee. Doubly resonant metallic nanostructure for high conversion efficiency of second harmonic generation. *Opt. Express*, 20(5):4856–4870, 2012.

[79] I.-Y. Park, S. Kim, J. Choi, D.-H. Lee, Y.-J. Kim, M. F. Kling, M. I. Stockman, and S.-W. Kim. Plasmonic generation of ultrashort extreme-ultraviolet light pulses. *Nat Photon*, 5(11):677–681, 2011.

[80] N. Pfullmann, C. Waltermann, M. Kovačev, V. Knittel, R. Bratschitsch, D. Akemeier, A. Hütten, A. Leitenstorfer, and U. Morgner. Nano-antenna-assisted harmonic generation. *Appl. Phys. B*, 113(1):75–79, 2013.

[81] Y.-Y. Yang, A. Scrinzi, A. Husakou, Q.-G. Li, S. L. Stebbings, F. Süßmann, H.-J. Yu, S. Kim, E. Rühl, J. Herrmann, X.-C. Lin, and M. F. Kling. High-harmonic and single attosecond pulse generation using plasmonic field enhancement in ordered arrays of gold nanoparticles with chirped laser pulses. *Opt. Express*, 21(2):2195–2205, 2013.

[82] T. Shaaran, M. F. Ciappina, and M. Lewenstein. Quantum-orbit analysis of high-order-harmonic generation by resonant plasmon field enhancement. *Phys. Rev. A*, 86(2):023408, 2012.

[83] M. F. Ciappina, J. Biegert, R. Quidant, and M. Lewenstein. High-order-harmonic generation from inhomogeneous fields. *Phys. Rev. A*, 85(3):033828, 2012.

[84] B. C. Stuart, M. D. Feit, S. Herman, A. M. Rubenchik, B. W. Shore, and M. D. Perry. Nanosecond-to-femtosecond laser-induced breakdown in dielectrics. *Phys. Rev. B*, 53(4):1749–1761, 1996.

[85] M. Lenzner, J. Krüger, S. Sartania, Z. Cheng, C. Spielmann, G. Mourou, W. Kautek, and F. Krausz. Femtosecond optical breakdown in dielectrics. *Phys. Rev. Lett.*, 80(18):4076–4079, 1998.

[86] S. I. Anisimov, B. L. Kapeliov, and T. L. Perelman. Electron-Emission From Surface of Metals Induced By Ultrashort Laser Pulses. *Zh. Eksp. Teor. Fiz.*, 66(2):776–781, 1974.

[87] J. König, S. Nolte, and A. Tünnermann. Plasma evolution during metal ablation with ultrashort laser pulses. *Opt. Express*, 13(26):10597–10607, 2005.

[88] D. von der Linde, K. Sokolowski-Tinten, and J. Bialkowski. Laser-solid interaction in the femtosecond time regime. *Appl. Surf. Sci.*, 109:1–10, 1997.

[89] S.-S Wellershoff, J. Hohlfeld, J. Güdde, and E. Matthias. The role of electron–phonon coupling in femtosecond laser damage of metals. *Appl. Phys. A*, 69(1):S99–S107, 1999.

[90] P. B. Corkum, F. Brunel, N. K. Sherman, and T. Srinivasan-Rao. Thermal Response of Metals to Ultrashort-Pulse Laser Excitation. *Phys. Rev. Lett.*, 61(25):2886–2889, 1988.

[91] B. N. Chichkov, C. Momma, S. Nolte, F. von Alvensleben, and A. Tünnermann. Femtosecond, picosecond and nanosecond laser ablation of solids. *Appl. Phys. A*, 63(2):109–115, 1996.

[92] D. Ashkenasi, M. Lorenz, R. Stoian, and A. Rosenfeld. Surface damage threshold and structuring of dielectrics using femtosecond laser pulses: the role of incubation. *Appl. Surf. Sci.*, 150(1-4):101–106, 1999.

[93] B. C. Stuart, M. D. Feit, S. Herman, A. M. Rubenchik, B. W. Shore, and M. D. Perry. Optical ablation by high-power short-pulse lasers. *J. Opt. Soc. Am. B*, 13(2):459–468, 1996.

[94] Y. Jee, M. F. Becker, and R. M. Walser. Laser-induced damage on single-crystal metal surfaces. *J. Opt. Soc. Am. B*, 5(3):648–659, 1988.

[95] J. Güdde, J. Hohlfeld, J. G. Müller, and E. Matthias. Damage threshold dependence on electron–phonon coupling in Au and Ni films. *Appl. Surf. Sci.*, 127–129(0):40–45, 1998.

[96] J. Bonse, J. M. Wrobel, J. Krüger, and W. Kautek. Ultrashort-pulse laser ablation of indium phosphide in air. *Appl. Phys. A*, 72(1):89–94, 2001.

[97] X. Ni, C.-Y. Wang, Li Yang, J. Li, L. Chai, W. Jia, R. Zhang, and Z. Zhang. Parametric study on femtosecond laser pulse ablation of Au films. *Appl. Surf. Sci.*, 253(3):1616–1619, 2006.

[98] J. M. Liu. Simple technique for measurements of pulsed Gaussian-beam spot sizes. *Opt. Lett.*, 7(5):196–198, 1982.

[99] J. Krüger, D. Dufft, R. Koter, and A. Hertwig. Femtosecond laser-induced damage of gold films: Photon-Assisted Synthesis and Processing of Functional Materials - E-MRS-H Symposium. *Appl. Surf. Sci.*, 253(19):7815–7819, 2007.

[100] M. Zürch. *Limitations of Ultra-Fast Nonlinear Optics in Nanostructured Samples.* Diplomarbeit, Friedrich-Schiller-Universität, Jena, 2010.

[101] C. de Marco, S. M. Eaton, R. Suriano, S. Turri, M. Levi, R. Ramponi, G. Cerullo, and R. Osellame. Surface Properties of Femtosecond Laser Ablated PMMA. *ACS Appl. Mater. Interfaces*, 2(8):2377–2384, 2010.

[102] D. Cialla, R. Siebert, U. Hübner, R. Möller, H. Schneidewind, R. Mattheis, J. Petschulat, A. Tünnermann, T. Pertsch, B. Dietzek, and J. Popp. Ultrafast plasmon dynamics and evanescent field distribution of reproducible surface-enhanced Raman-scattering substrates. *Anal. Bioanal. Chem.*, 394(7):1811–1818, 2009.

[103] A. Plech, V. Kotaidis, M. Lorenc, and J. Boneberg. Femtosecond laser near-field ablation from gold nanoparticles. *Nat Phys*, 2(1):44–47, 2006.

[104] A. Plech, P. Leiderer, and J. Boneberg. Femtosecond laser near field ablation. *Laser & Photon. Rev.*, 3(5):435–451, 2009.

[105] V. K. Valev, D. Denkova, X. Zheng, A. I. Kuznetsov, C. Reinhardt, B. N. Chichkov, G. Tsutsumanova, E. J. Osley, V. Petkov, B. de Clercq, A. V. Silhanek, Y. Jeyaram, V. Volskiy, P. A. Warburton, G. A. E. Vandenbosch, S. Russev, O. A. Aktsipetrov, M. Ameloot, V. V. Moshchalkov, and T. Verbiest. Plasmon-Enhanced Sub-Wavelength Laser Ablation: Plasmonic Nanojets. *Advanced Materials*, 24(10):OP29–OP35, 2012.

[106] A. Habenicht, M. Olapinski, F. Burmeister, P. Leiderer, and J. Boneberg. Jumping Nanodroplets. *Science*, 309(5743):2043–2045, 2005.

[107] M. J. Weber. *Handbook of optical materials.* CRC Press, Boca Raton, 2003.

[108] G. Xu, Y. Chen, M. Tazawa, and P. Jin. Influence of dielectric properties of a substrate upon plasmon resonance spectrum of supported Ag nanoparticles. *Appl. Phys. Lett*, 88(4):043114, 2006.

[109] J. F. Nye and M. V. Berry. Dislocations in Wave Trains. *Proc. R. Soc. London Ser. A*, 336(1605):165–190, 1974.

[110] M. V. Berry. Making waves in physics. *Nature*, 403(6765):21, 2000.

[111] D. Rozas, C. T. Law, and G. A. Jr. Swartzlander. Propagation dynamics of optical vortices. *J. Opt. Soc. Am. B*, 14(11):3054–3065, 1997.

[112] Z. S. Sacks, D. Rozas, and G. A. Jr. Swartzlander. Holographic formation of optical-vortex filaments. *J. Opt. Soc. Am. B*, 15(8):2226–2234, 1998.

[113] M. W. Beijersbergen, R. P. C. Coerwinkel, M. Kristensen, and J. P. Woerdman. Helical-wavefront laser beams produced with a spiral phaseplate. *Opt. Commun.*, 112(5–6):321–327, 1994.

[114] A. G. Peele, P. J. McMahon, D. Paterson, C. Q. Tran, A. P. Mancuso, K. A. Nugent, J. P. Hayes, E. Harvey, B. Lai, and I. McNulty. Observation of an x-ray vortex. *Opt. Lett.*, 27(20):1752–1754, 2002.

[115] N. R. Heckenberg, R. McDuff, C. P. Smith, and A. G. White. Generation of optical phase singularities by computer-generated holograms. *Opt. Lett.*, 17(3):221–223, 1992.

[116] K. Bezuhanov, A. Dreischuh, G. G. Paulus, M. G. Schätzel, H. Walther, D. N. Neshev, Wieslaw Królikowski, and Y. S. Kivshar. Spatial phase dislocations in femtosecond laser pulses. *J. Opt. Soc. Am. B*, 23(1):26–35, 2006.

[117] M. W. Beijersbergen, L. Allen, H. E. L. O. van der Veen, and J. P. Woerdman. Astigmatic laser mode converters and transfer of orbital angular momentum. *Opt. Commun.*, 96(1–3):123–132, 1993.

[118] L. Allen, M. W. Beijersbergen, R. J. C. Spreeuw, and J. P. Woerdman. Orbital angular momentum of light and the transformation of Laguerre-Gaussian laser modes. *Phys. Rev. A*, 45(11):8185, 1992.

[119] J. H. Poynting. The Wave Motion of a Revolving Shaft, and a Suggestion as to the Angular Momentum in a Beam of Circularly Polarised Light. *Proc. R. Soc. London Ser. A*, 82(557):560–567, 1909.

[120] R. A. Beth. Mechanical Detection and Measurement of the Angular Momentum of Light. *Physical Review*, 50(2):115–125, 1936.

[121] S. M. Barnett and L. Allen. Orbital angular momentum and nonparaxial light beams. *Opt. Commun.*, 110(5-6):670–678, 1994.

[122] S. J. van Enk and G. Nienhuis. Eigenfunction description of laser beams and orbital angular momentum of light. *Opt. Commun.*, 94(1–3):147–158, 1992.

[123] J. E. Curtis and D. G. Grier. Structure of Optical Vortices. *Phys. Rev. Lett.*, 90(13):133901, 2003.

[124] D. G. Grier. A revolution in optical manipulation. *Nature*, 424(6950):810–816, 2003.

[125] A. Mair, A. Vaziri, G. Weihs, and A. Zeilinger. Entanglement of the orbital angular momentum states of photons. *Nature*, 412(6844):313–316, 2001.

[126] G. Molina-Terriza, J. P. Torres, and L. Torner. Twisted photons. *Nat Phys*, 3(5):305–310, 2007.

[127] S. J. van Enk. Selection rules and centre-of-mass motion of ultracold atoms. *Quantum Opt.*, 6(5):445, 1994.

[128] A. Picón, A. Benseny, J. Mompart, J. R. de Vázquez Aldana, L. Plaja, G. F. Calvo, and L. Roso. Transferring orbital and spin angular momenta of light to atoms. *New J. Phys.*, 12(8):083053, 2010.

[129] A. Picón, J. Mompart, J. R. de Vázquez Aldana, L. Plaja, G. F. Calvo, and L. Roso. Photoionization with orbital angular momentum beams. *Opt. Express*, 18(4):3660–3671, 2010.

[130] A. G. Peele, K. A. Nugent, A. P. Mancuso, D. Paterson, I. McNulty, and J. P. Hayes. X-ray phase vortices: theory and experiment. *J. Opt. Soc. Am. A*, 21(8):1575–1584, 2004.

[131] E. Hemsing, A. Knyazik, M. Dunning, D. Xiang, A. Marinelli, C. Hast, and J. B. Rosenzweig. Coherent optical vortices from relativistic electron beams. *Nat Phys*, 9(9):549–553, 2013.

[132] J. Courtial, K. Dholakia, L. Allen, and M. J. Padgett. Second-harmonic generation and the conservation of orbital angular momentum with high-order Laguerre-Gaussian modes. *Phys. Rev. A*, 56(5):4193–4196, 1997.

[133] A. Berzanskis, A. Matijosius, A. Piskarskas, V. Smilgevicius, and A. Stabinis. Conversion of topological charge of optical vortices in a parametric frequency converter. *Opt. Commun.*, 140(4–6):273–276, 1997.

[134] A. V. Gorbach and D. V. Skryabin. Cascaded Generation of Multiply Charged Optical Vortices and Spatiotemporal Helical Beams in a Raman Medium. *Phys. Rev. Lett.*, 98(24):243601, 2007.

[135] D.-S Ding, Z.-Y Zhou, B.-S Shi, X.-B Zou, and G.-C Guo. Linear up-conversion of orbital angular momentum. *Opt. Lett.*, 37(15):3270–3272, 2012.

[136] J. Strohaber, M. Zhi, A. V. Sokolov, A. A. Kolomenskii, G. G. Paulus, and H. A. Schuessler. Coherent transfer of optical orbital angular momentum in multi-order Raman sideband generation. *Opt. Lett.*, 37(16):3411–3413, 2012.

[137] D. N. Neshev, A. Dreischuh, G. Maleshkov, M. Samoc, and Y. S. Kivshar. Supercontinuum generation with optical vortices. *Opt. Express*, 18(17):18368–18373, 2010.

[138] W. J. Firth and D. V. Skryabin. Optical Solitons Carrying Orbital Angular Momentum. *Phys. Rev. Lett.*, 79(13):2450–2453, 1997.

[139] A. Dreischuh, G. G. Paulus, F. Zacher, F. Grasbon, D. N. Neshev, and H. Walther. Modulational instability of multiple-charged optical vortex solitons under saturation of the nonlinearity. *Phys. Rev. E*, 60(6):7518, 1999.

[140] S. Eyring. *Extremely Nonlinear Optics with wavefront controlled ultra-short laser pulses.* Dissertation, Bayerische Julius-Maximilians-Universität, Würzburg, 2011.

[141] M. Zürch, C. Kern, P. Hansinger, A. Dreischuh, and C. Spielmann. Strong-field physics with singular light beams. *Nat Phys*, 8(10):743–746, 2012.

[142] P. Polynkin, C. Ament, and J. V. Moloney. Self-Focusing of Ultraintense Femtosecond Optical Vortices in Air. *Phys. Rev. Lett.*, 111(2):023901, 2013.

[143] A. Vinçotte and L. Bergé. Femtosecond Optical Vortices in Air. *Phys. Rev. Lett.*, 95(19):193901, 2005.

[144] A. G. Peele and K. A. Nugent. X-ray vortex beams: A theoretical analysis. *Opt. Express*, 11(19):2315–2322, 2003.

[145] C. Kern. *Measuring low-order aberrations in High Harmonic beams with a Hartmann wave front sensor.* Diplomarbeit, Bayerische Julius-Maximilians-Universität, Würzburg, 2008.

[146] D. L. Fried. Adaptive optics wave function reconstruction and phase unwrapping when branch points are present. *Opt. Commun.*, 200(1-6):43–72, 2001.

[147] F. A. Starikov, G. G. Kochemasov, S. M. Kulikov, A. N. Manachinsky, N. V. Maslov, A. V. Ogorodnikov, S. A. Sukharev, V. P. Aksenov, I. V. Izmailov, F. Y. Kanev, V. V. Atuchin, and I. S. Soldatenkov. Wavefront reconstruction of an optical vortex by a Hartmann-Shack sensor. *Opt. Lett.*, 32(16):2291–2293, 2007.

[148] K. Murphy, D. Burke, N. Devaney, and C. Dainty. Experimental detection of optical vortices with a Shack-Hartmann wavefront sensor. *Opt. Express*, 18(15):15448–15460, 2010.

[149] G. Indebetouw. Optical Vortices and Their Propagation. *J. Mod. Opt.*, 40(1):73–87, 1993.

[150] Y. V. Izdebskaya, J. Rebling, A. S. Desyatnikov, and Y. S. Kivshar. Observation of vector solitons with hidden vorticity. *Opt. Lett.*, 37(5):767–769, 2012.

[151] C. Hernández-García, A. Picón, J. San Román, and L. Plaja. Attosecond Extreme Ultraviolet Vortices from High-Order Harmonic Generation. *Phys. Rev. Lett.*, 111(8):083602, 2013.

[152] P. B. Corkum. Private communication.

List of Figures

List of Tables

Appendix A

The Laser Source and Shot Delivery System

For damage measurements in chapter 3 and HHG experiments in chapter 6, we employed a femtosecond titanium:sapphire (Ti:Sa) amplified laser. The system is a FEMTOPOWER *compact pro*, which is seeded by a Ti:Sa oscillator delivering pulses of a few nanojoules at 80 MHz repetition rate. The spectrum of slightly above 100 nm FWHM bandwidth around the 800 nm central wavelength supports pulses of $\sim$ 8 fs duration. The 9-pass chirped-pulse amplifier is pumped by a Q-switched, frequency-doubled 527 nm kHz neodymium:yttrium-lithium-fluoride (Nd:YLF) laser. After stretching, amplification and re-compression, the output consists of pulses of up to 0.8 mJ in energy, clocked by the pump to about 1 kHz. The bandwidth, due to gain-narrowing in the amplifier, is left at $\gtrsim$ 50 nm, allowing for pulses of slightly less than 30 fs duration. Figure A.1 **a** summarises the specifications of the system.

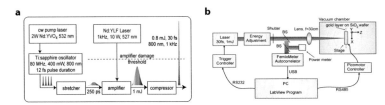

Figure A.1: a, Principle and specifications of the laser system (from [145]) and **b**, Schematic of the pulse delivery for LID measurements used for this work (from [100]).

A number of self-built devices are used to control the number and energy of laser pulses for Laser-induced Damage measurements (for a detailed description, see [100]). A large attenuation and rough tuning of the pulse energy regime is obtained by using up to three metallic neutral density (ND) filters in front of the pulse diagnostic and eventual experiment. The fine adjustment is achieved by an additional polarisation optic in the compressor part of the laser. We place an adjustable wave-plate before the prisms of the pulse compressor. The p-polarised output of the amplifier hits a total of 16 refracting surfaces before leaving the laser as a compressed pulse. All prisms are employed in a Brewster's angle configuration, such that reflections of the beam on the compressor optics are kept to a minimum. By rotating the input polarisation a certain amount away from the loss-less p-state, the 16th power of the reflected fraction is removed from the transmitted beam. The combined prism surfaces thus act as an analysing polariser, and the output beam is close to p-polarisation again. In [100], the exact calibration of output laser power vs. half-wave-plate rotation is presented. It is also documented that there is no significant effect of this technique on the spectrum and thus the achieved pulse length of the femtosecond laser. Without ND filters, every power value from full output down to the ASE level of the pumped amplifier crystal can be selected at a precision of less than a few mW.

To control the applied number of pulses, we electronically access the pulse picking procedure of the free-running laser. The amplification of kHz-clocked pulses in the original laser is achieved by picking one in approximately 80,000 pulses and rotating its polarisation by 90° after four passes through the amplifying crystal. This is done by a fast switching Pockel's cell. In synchronisation with the kHz Q-switched pump, the pulse overlapping with the highest inversion inside the pumped crystal is chosen and allowed to be transmitted for the five remaining passes. For our purposes, an electronic device to control the triggering signal was constructed. This works such that the Pockel's cell is left without an impulse-giving signal if no output is desired. In this state, the cell does not act as a polarisation rotator at all, so all pulses are dumped by the filtering beam-splitting cubes. However, with the use of a 'fire'-button, the signal can immediately be restored, and normal laser operation is reached after a couple of milliseconds. Alternatively, an electronically counted number of shots can be specified and allowed to be amplified and leave the laser. The remaining output from the laser when the Pockel's cell is untriggered, given again by the ASE threshold, can be crudely blocked by a mechanical shutter before any actual experiment.

In [100], electronic details of the employed circuits are given. Also, there is described a small correction for output power that is due to the build-up of the Pockel's cell to full functionality. As was said, full performance of the polarisation switching is achieved only after a few milliseconds, so the first couple of shots after triggering the output will contain less energy than those in free-running operation. As reliable on-site powers could only be determined with the full repetition rate, the pulse energies applied in experiments with small shot numbers are multiplied by a factor below but close to 1.

Especially for the intensity regime that proved to be relevant for the characterisation of LID on solid layers, it was important to be able to perform these experiments in a rough ($\lesssim 10^{-2}$ mbar) vacuum. This is to exclude effects of ionised air attacking the surface, which would muddle up the effect of actual surface material damage. Inside the vacuum chamber, an externally controlled stage system can comfortably move the samples under test in lateral x and y directions, as well as scan the sample with relation to the focus in the z dimension. The focus is obtained in the same geometry as is usually used in the HHG setup inside the same chamber, by use of a lens ($f = 35$ cm, $f/\# \approx 20$). This results in spot diameters on the target of ~ 40–$60\,\mu$m. The described elements and their constellation in the setup for LID measurements are shown in Fig. A.1 **b**.

Appendix B

Design of the Bow-tie Sample

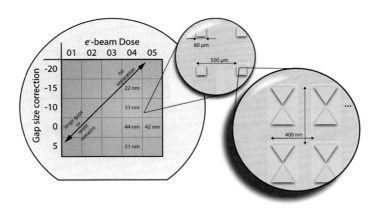

Figure B.1: Schematic of the bow-tie sample.

The truncated grey circle (left) represents the substrate, the square grid is the processed area. The grid is orthogonally indexed by the two production parameters, gap size correction and e^--beam dose. Due to reasons explained in the body text of section 3.3, most parameter configurations produce undesired results in the respective sections, marked by the red shading. The diagonal arrow illustrates the trends in the resulting structures. Beneficial regions are shaded green, and evaluated average gap sizes are written out in each square where the resist was fully removed, and the triangles non-touching. All parameter configurations consist of 11×11 identical arrays of 150×150 bow-ties. Size and distance of the arrays, as well as the periodicity of single antennas, are illustrated in the respective 'magnified' insets.

Upon delivery, the sample was characterised by SEM, showing typical antenna results in all procession regimes. Some examples are shown in Fig. B.2.

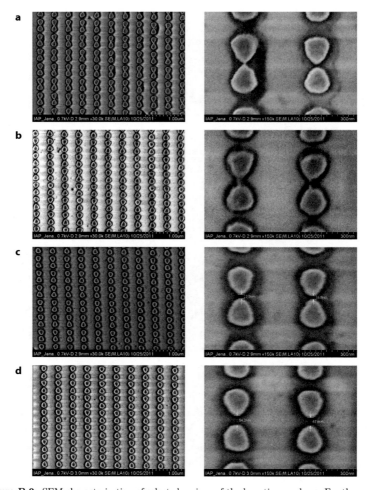

Figure B.2: SEM characterisation of selected regions of the bow-tie sample. **a,** For the smallest gap size correction G and the largest e^--beam dose Dos, most triangles are not fully separated. **b,** For the three smallest values of Dos, some resist remains between the triangles. For fully featured bow-ties ('$G-15$, $Dos04$' (**c**) and '$G+5$, $Dos04$' (**d**)), correct choice of parameter G determines whether the gaps form satisfactorily (**c**) or lead to probably decoupled single triangles (**d**) (Images courtesy of M. Banasch).

Appendix C

Details of the Vortex Generation and HHG Setup and Experiments

The employed Spatial Light Modulator (SLM) is a model X10468-02 (800 ± 50 nm) by Hamamatsu. It works as a high-reflective component with an additional top layer consisting of a nematic liquid crystal array. At near SVGA spatial resolution (792 × 600 pixels), a specific area of pixel size $20 \times 20\,\mu\mathrm{m}^2$ can be addressed to act on the respective beam fragment as a material of variable refractive index. This is achieved by applying a voltage to the liquid crystal at the site of the pixel, encoded in the black-white value of an image from a conventional computer graphics adapter. Apart from an arbitrary global offset, 'black' corresponds to no additional phase shift, while 'white' corresponds to an approximate 2π phase shift. The full 2π value depends on the absolute wavelength illuminating the SLM. Greater continuous shifts can be realised by wrapping of the phase, i.e., rejoining the neighbouring pixels with $\phi \equiv \phi + n \cdot 2\pi$ steering information. The resolution of modulation is 8-bit (256). Some further details on the addressing and controlling of the device are given in [140].

An illustration of the beamline relevant for the experiments presented in chapter 6 is shown in Fig. C.1. While sub-figure **a** represents the conceptual insertion of the SLM as described above, Figs. C.1 **b** and **c** show the generation chamber and the two subsequent optional detection schemes. After HHG and fundamental filtering by the usual aluminium filters, either the freely propagating XUV beam profile can be viewed by a CCD (Andor iKon-L) with a large area chip, or optionally the interaction region in the generation chamber is imaged onto the entrance slit of an XUV spectrometer (McPherson 248/310G). The spatial observation setup can accommodate additional phase examination techniques, described in more detail in the respective paragraph of section 6.3.

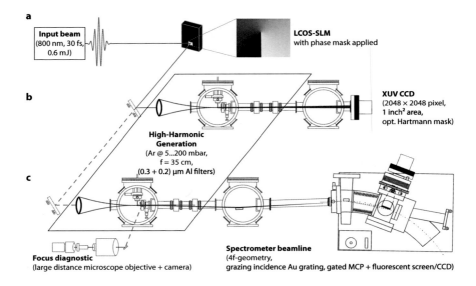

Figure C.1: Beamline setups used for HHG with a fundamental beam carrying an Optical Vortex. **a**, OV generation scheme as described in section 6.1, using the LCOS-SLM as a reflective component in the beam delivery. **b**, Detection and evaluation of the spatial profile of the High-Harmonic beam. The large-area XUV CCD can be equipped with a Hartmann hole mask, as described in [145]. **c**, Optional switching of the hindmost chamber with another, equipped with a toroidal mirror, allows to image the generation region onto an XUV spectrometer. The HHG setup, with the specifications as given in the graphic, is the same for **b** and **c**.

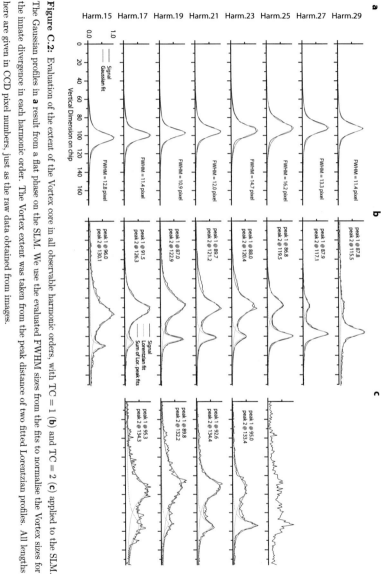

Figure C.2: Evaluation of the extent of the Vortex core in all observable harmonic orders, with TC = 1 (**b**) and TC = 2 (**c**) applied to the SLM. The Gaussian profiles in **a** result from a flat phase on the SLM. We use the evaluated FWHM sizes from the fits to normalise the Vortex sizes for the innate divergence in each harmonic order. The Vortex extent was taken from the peak distance of two fitted Lorentzian profiles. All lengths here are given in CCD pixel numbers, just as the raw data obtained from images.